Food Science Text Series

The Food Science Text Series provides faculty with the leading teaching tools. The Editorial Board has outlined the most appropriate and complete content for each food science course in a typical food science program and has identified textbooks of the highest quality, written by the leading food science educators.

For further volumes:
http://www.springer.com/series/5999

Rotimi E. Aluko

Functional Foods
and Nutraceuticals

 Springer

Rotimi E. Aluko
Department of Human Nutritional Sciences
University of Manitoba
Winnipeg, MB, Canada

ISSN 1572-0330
ISBN 978-1-4939-5064-5   ISBN 978-1-4614-3480-1 (eBook)
DOI 10.1007/978-1-4614-3480-1
Springer New York Dordrecht Heidelberg London

Printed on acid-free paper

Springer is part of Springer Science+Business Media (www.springer.com)

Dedicated to my wife, Rita, and our children, Victor and Rachael

General Introduction

History: It is an established fact that foods provide nutrients that nourish our body and keep our system in proper working conditions. However, from early civilization it was also known that certain foods confer additional health benefits to human beings such as prevention and treatment of various types of diseases. "Let food be thy medicine and let your medicine be your food" is a popular quote from Hippocrates (460–370 BC) that emphasizes the role of foods in disease prevention and recognizes a separate role for food in addition to being nutrient providers. Recently, scientists have become focused on the health-promoting effects of foods and there is now abundance evidence that support the role of various foods and their components in promoting human health. In 1989 the word "nutraceutical," a blend of "nutrition" and "pharmaceutical" was coined by Dr. Stephen De Felice, a physician who founded the Foundation for Innovation in Medicine, USA. At the time, Dr. De Felice defined "nutraceutical" as "any food or parts of a food that provides medical or health benefits, including the prevention and treatment of diseases". Since this initial definition, the term "functional foods" has also been added to link consumption of certain foods or food products with disease prevention and improved health benefits. Development and regulatory oversight of functional foods began in earnest in Japan in the early eighties with advances in chemical identification of bioactive compounds, processing and formulation of foods as well as elucidation of molecular mechanisms involved in the modulation of metabolic disorders. The initial regulatory environment for functional foods was established by Japan in 1991 with the introduction of "foods for specified health use" (FOSHU) policy that enabled production and marketing of health-promoting foods. Since 1991 over 600 FOSHU products are now available in the Japanese market. The initiative in Japan has spurred a rapid growth in the global functional foods market especially in the USA, European Union, and Canada, all of which now have various regulatory bodies to govern the manufacture and marketing of health-promoting food products. The availability of regional regulatory bodies has spurred intense global research and development aimed at identifying new bioactive compounds that could be used to formulate functional foods and nutraceuticals. While the potential therapeutic activities of several compounds have been reported, there is still paucity of information regarding the molecular mechanisms of action. Most of what is known about the role of bioactive natural compounds in human health has arisen mainly from in vitro and animal experiments, though human intervention trials are also occurring.

Definitions: These health-promoting foods or compounds are generally classified into two major categories: (1) Functional foods are in fact products that may look like or be a conventional food and be consumed as part of a usual diet, but apart from supplying nutrients they can reduce the risk of chronic diseases such as cancer, hypertension, kidney malfunction, etc. A typical example of a functional food is tomato fruit which is packed with a specific type of compound that helps to remove toxic compounds from our body and thereby prevent damage to essential organs like the heart, kidney, lungs, brain, etc. Other typical examples of functional foods include soybean, fish, oat meal, cereal bran (wheat, rice), and tea (green and black). Apart from traditional foods, there are also functional foods that are produced through food processing such as the antihypertensive sour milk that has been shown to reduce blood pressure in human beings. (2) Nutraceuticals are health-promoting compounds or products that have been isolated or purified from food sources and they are generally sold in a medicinal (usually pill) form. A good example is a group of compounds called isoflavones that are isolated from soybean seeds and packaged into pills that women can use instead of synthetic compounds during hormone replacement therapy. Other examples of nutraceutical products include fish oil capsules, herb extracts, glucosamine and chondroitin sulfate pills, lutein-containing multivitamin tablets, and antihypertensive pills that contain fish protein-derived peptides.

The content of this book has been organized based on two main sections; the first describes the bioactive properties of major nutrients (carbohydrates, proteins, lipids, and polyphenols) while the second discusses the role of major food types (soybean, fish, milk, fruits, and vegetables, and miscellaneous foods) in health promotion. It is hoped that users of this book will benefit from information provided on the potential mechanisms that have been proposed for the bioactivity of various foods and their components.

Contents

Part I

Nutrient Components of Foods

Bioactive Carbohydrates

1.1 Introduction

Carbohydrates are important sources of energy in our diet, but certain structural characteristics enable their use beyond basic nutrition. The most important structural feature of bioactive carbohydrates is resistance to digestion in the upper tract of the gastrointestinal tract (GIT), primary because of the presence of glycosidic bonds that are different from the digestive enzyme-susceptible α-1,4 and α-1,6 linkages. Carbohydrates that are not digested in the upper tract of the GIT will reach the colon and become food for the microflora that converts them into bioactive compounds. Or the carbohydrate is used as fuel by beneficial bacteria in order to grow and multiply at the expense of pathogenic microorganisms.

1.2 Trehalose [α-D-glucopyranosyl-(1→1)-α-D-glucopyranoside]

This disaccharide (Fig. 1.1) is a nonreducing sugar consisting of two molecules of D-glucose joined together by α-1,1 bond. Because the two glucose units are joined together through their respective carbonyl carbon atoms, the disaccharide has nonreducing properties and cannot participate in Maillard reactions. Trehalose can be found in sunflower seeds, plants belonging to the *Selaginella* family, some mushrooms (e.g., Shiitake and Judas's ear), and baker's yeast. Trehalose can be used as a sweetener, but important physiological benefits of dietary trehalose have been reported. In the small intestine, trehalose is broken down into the two component D-glucose residues by the enzyme trehalase; therefore, the sugar is digestible. Biological effects associated with trehalose include reduced insulinemia when compared to glucose, attenuation of adipocyte hypertrophy, inhibition of bone resorption, and suppression of inflammatory response. Specifically trehalose has been shown to reduce plasma insulin levels during oral glucose tolerance tests. Therefore, trehalose may be used as an agent to protect against metabolic syndrome because of its ability to reduce insulin secretion and downregulate expression of monocyte chemoattractant protein-1 (MCP-1), an inflammatory compound.

1.3 Polysaccharides

These are complex polymers of various monosaccharides such as hexoses, pentoses, and their acids. Polysaccharides that are of great interest in human health are those that are resistant to digestion within the upper intestinal tract and are often referred to as dietary fiber. Such polysaccharides pass through the colon undigested or partly digested but are then subjected to microbial fermentation once they reach the colon. The health benefits of dietary fiber arise from one main event in the upper GIT and two main events that are associated with the metabolic fate in the colon. In the upper GIT, dietary fiber dilutes caloric content of foods, and soluble fibers enhance viscosity

R.E. Aluko, *Functional Foods and Nutraceuticals*, Food Science Text Series,
DOI 10.1007/978-1-4614-3480-1_1, © Springer Science+Business Media, LLC 2012

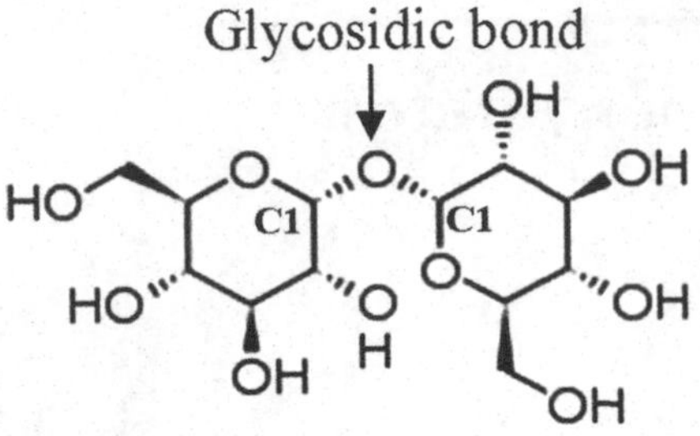

Glycosidic bond

Fig. 1.1 Chemical structure of trehalose showing the C1→C1 glycosidic bond and position of the glycosidic bond

of luminal contents, which leads to decrease nutrient absorption and concomitant increase in fecal weight. In the colon, firstly, the products arising from microbial fermentation have bioactive properties either locally within the colon or systemically as a result of absorption into the blood followed by distribution to target organs. Secondly, the growth of certain microorganisms (mostly lactic acid bacteria) is greatly enhanced by the dietary fibers that serve as food for the microbes; usually this microbial growth occurs at the expense of pathogenic microbes that are then displaced from the colon surface and expelled in fecal matter. Thus, in general, increase consumption of fiber is associated with reduced prevalence of cancer and coronary heart disease. The mechanisms involved in these health-promoting benefits of dietary fiber are believed to involve carcinogen-binding, antioxidant properties, production of short-chain fatty acids (SCFAs), reduced calorie density of foods, and increased bile (cholesterol) excretion. Dietary fiber is made up of these undigested polysaccharides, which include cellulose, hemicellulose, beta-glucans, lignin, pectins, and gums. Traditionally, the following reasons have been used for emphasizing higher intake of dietary fiber as a means of promoting human health:

(a) *Bowel health*: It refers to the ability to maintain an overwhelming mass of beneficial microorganisms in the colon and also the ability to reduce the level and transient time of toxins (especially carcinogens) in the digestive tract. Increased growth of beneficial lactic acid bacteria is promoted, which leads to increased colon acidity, and the toxic effect reduces number of pathogenic microbes.

(b) *Healthy weight*: Dietary fiber reduces the caloric density of foods by physical dilution and by entrapment of nutrients, which reduces the rate of digestion and absorption (bioavailability) of nutrients, especially glucose and lipids. Soluble fibers that form viscous hydrated masses suppress appetite better than the less viscous fibers. This is due to the large amount within the fiber that increases stomach distension and triggers the feeling of fullness. The high viscosity prolongs presence of nutrients in the small intestine, downregulates release of appetite-stimulating hormones, and leads to reduced appetite. On a long-term basis, it is believed that short-chain fatty acids produced during fermentation of fibers in the colon have appetite-suppressing effect, especially by acting on the brain.

(c) *Heart health*: Dietary fiber promotes increased fecal bulk, which helps to trap and remove bile acids and cholesterols from the intestinal tract. By reducing rate of reabsorption of cholesterol, dietary fibers help to lower plasma levels of LDL cholesterol and triglycerides, which reduces the risk of atherosclerosis and associated cardiovascular symptoms such as hypertension and stroke.

(d) *Cancer*: There is an inverse relationship between dietary fiber consumption and breast cancer. Fiber within the intestinal lumen retards reabsorption of estrogens, which increases fecal estrogen level but leads to lower levels of blood and urine estrogen. Wheat-bran-bound alkylresorcinols, a lipophilic compound, have been found to be toxic to prostate cancer cells and may be responsible for the cancer-preventive activity of the bran. Alkylresorcinol with a shorter chain (C17:0) was found to be more potent than those with longer chains when tested against the prostate cancer cells. Introduction of a double bond to long-chain (C19:1, C21:1, and C23:1) alkylresorcinols led to increased potency as anticancer agents.

(e) *Fetal development*: The typical western diet is rich in fats and may contribute to excessive oxidative stress, a known risk factor for improper fetal development such as embryonic growth retardation and toxicity, and even causes miscarriages. Moreover, oxidative damage may cause abnormal intrauterine development, which can instigate onset of metabolic diseases later in life. In female rats that consumed a high-fat diet, inclusion of high fiber (mostly as insoluble fiber) content led to reduced number of aborted fetuses when compared to the rats that consumed the high-fat, low-fiber diet. The high-fiber diet also ameliorated placenta malformation (especially necrosis) and led to about 20% increase in number of fetuses (litter size) when compared to the high-fat, low-fiber diet. Antioxidative stress was reduced significantly by the high-fiber diet as evident in the lower levels of malondialdehyde, coupled with higher superoxide anion and hydroxyl radical scavenging capacities in maternal serum and placenta. Expression of genes for various antioxidant enzymes such as glutathione peroxidase, hypoxia-inducible factor 1α, Cu, Zn-superoxide dismutase, and Mn-superoxide dismutase was upregulated by the high-fiber diet. Fiber fermentation in the colon leads to formation of SCFAs, which are known to reduce susceptibility of DNA to oxidative damage and also induce activity of glutathione S-transferase. Therefore, consumption of high fiber as part of a high-fat diet could reduce oxidative stress by regulating mRNA expressions of antioxidant-related genes, and improvement of placenta health, fetal growth, and development.

General functions of fiber-stimulated intestinal microflora
- Energy salvage (lactose digestion, short-chain fatty acid production)
- Modulation of cell growth and differentiation
- Antagonism against pathogens
- Immune stimulation of the gut-associated lymphoid tissue
- Natural immunity against infections
- Production of vitamins
- Reduction of blood lipids
- Hydrolysis of insoluble fibers to release bioactive conjugated polyphenolic compounds

1.4 Soluble Fibers

In general, soluble dietary fibers mix very well with water to form highly hydrated masses that are almost completely fermented in the colon by microorganisms. Due to the high degree of breakdown by colon microflora, very little soluble fiber is excreted in the feces. But because the fiber supports increased growth of bacteria, dietary soluble fiber enhances stool weight because of the increase in bacteria mass. Apart from natural soluble fibers discussed below, the beneficial effects of chemically modified cellulose (e.g., methylcellulose and hydroxypropylmethylcellulose) in reducing postprandial glucose and plasma lipids have been attributed to increased solubility properties. While natural cellulose is mostly insoluble and produces no consistent physiological effects, chemical modification can reduce degree of crystallinity and increase amorphous components. The more amorphous chemically modified cellulose interacts better with water and forms a more viscous product than the native unmodified cellulose.

1.4.1 Pectin

This is a heteropolymer (more than one type of monosaccharide unit) consisting of a backbone of linear galacturonic acids linked by α-1,4 bonds and substituted with α-1,2 rhamnopyranose units that contain side chains of neutral sugars such as glucose, mannose, xylose, and galactose (Fig. 1.2). As expected, pectin is largely undigested in the upper GIT but is easily broken down through fermentation by colon microorganisms. Most abundant sources of pectin are citrus fruits that contain 0.5–3.5% by weight and located mostly in the peel. The main physiological effects of pectin are related to improved plasma glucose, cholesterol, and total lipid profiles. Therefore,

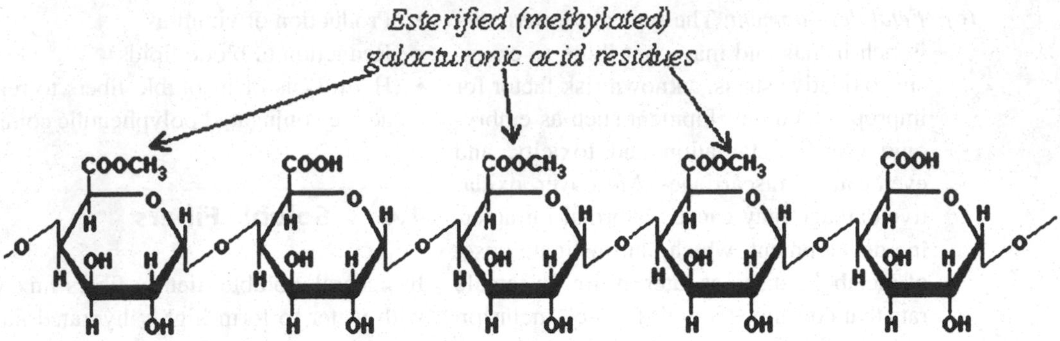

Esterified (methylated)
galacturonic acid residues

Fig. 1.2 Chemical structure of a typical repeating unit of pectin indicating 60% degree of esterification (3 methyl groups out of 5 galacturonic acid units) and the $\alpha 1 \rightarrow 4$ glycosidic linkages

dietary pectin may help in the prevention of chronic diseases such as obesity, diabetes, atherosclerosis, and cancer. The mechanism involved in pectin action is related to ability to form a gel or thickened solution, which can trap nutrients (cholesterol, bile acids, glucose) and reduce their absorption from the GIT. In the colon, pectin is fermented mostly by acid-producing bacteria (*Bifidobacteria* and *Lactobacillus*), and the resultant increased acidity could be lethal to pathogenic microorganisms. For example, in children and infants that suffer from intestinal infections, oral administration of pectin led to significantly lower diarrhea intensity, which was associated with reduced numbers of pathogenic bacteria such as *Klebsiella*, *Proteus*, *Shigella*, *Citrobacter*, and *Salmonella*. The anticancer effect of pectin on cancer growth was related to cancer-cell-binding ability, which also decreases cell migration. It is believed that pectin also acts as an anticancer agent by inhibiting galectin-3, a carbohydrate-binding protein (lectin) that has been implicated in the pathogenesis of tumors. With regard to the cardiovascular benefits, pectin has been shown to increase permeability of fibrin and decrease fibrin tensile strength. Mechanism of action is thought to be due to the effect of acetate, the predominant fermentation by-product from microbial degradation of pectin in the colon. Acetate is absorbed from the colon into the blood circulatory system where it modulates fibrin architecture to increase permeability and reduce protein-protein interactions (strength). This is important because increased aggregation of fibrin is an important

risk factor for the development of atherosclerosis, stroke, and coronary heart disease.

1.4.2 Guar Gum

This is a viscous polysaccharide consisting of galactose and mannose that is extracted from seeds of *Cyamopsis tetragonolobus*, a drought-tolerant leguminous herb. Dietary guar gum has hypocholesterolemic effects in animals and humans, which makes it an ingredient of choice in the formulation of functional foods for treatment of cholesterol-related cardiovascular diseases. The basic mechanism behind the hypocholesterolemic effects of guar gum is the ability to reduce concentration of free cholesterol in the liver. Guar gum upregulates the nuclear expression of sterol regulatory element-binding protein 2 (SREBP2), which then upregulates hepatic LDL receptor (LDLr). High levels of LDLr enhance the ability of the liver to remove cholesterol from circulation with concomitant decrease in plasma cholesterol level.

In healthy men, guar gum can improve cardiovascular health by decreasing fasting blood glucose, cholesterol, triacylglycerol, systolic blood pressure, diastolic blood pressure, and plasminogen activator inhibitor-1 activity. Additional health benefits of guar gum consumption include increases in glucose sensitivity, adipose tissue glucose uptake, and urinary excretion of sodium and potassium. In insulin-dependent diabetics, guar gum provides health benefits such as

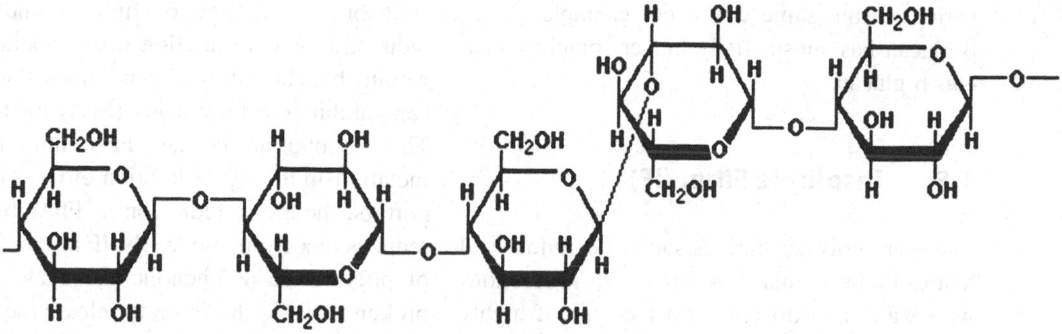

Fig. 1.3 Typical repeating unit of β-glucan showing the β1→4 and β1→3 glycosidic linkages, which make the polymer resistant to enzymatic breakdown in the gastrointestinal tract

decreased fasting blood glucose, glycosylated hemoglobin, and ratio of LDL cholesterol to HDL-cholesterol. Thus, dietary guar gum contributes to lowering the risk of developing atherosclerosis by enhancing cholesterol clearance from the blood circulatory system. Apart from the effect of guar gum at the molecular level in increasing expression of SREBP2, the highly viscous mass formed in the intestinal tract increases gastric emptying time and prolongs the intestinal absorption phase of fat, carbohydrates, and sodium to provide cholesterol-lowering effects. The increased gastric emptying time enhances satiety and could be of benefit to obese patients as a means of reducing caloric or nutrient intake.

1.4.3 Barley and Oat β-Glucan

Barley and oat endosperms contain soluble and highly viscous fiber in the form of β-glucan (Fig. 1.3), a linear polysaccharide consisting of glucose monomers joined together by β-1,4 and β-1,3 glycosidic linkages. The bioactive properties of β-glucan are due mainly to their effects on lipid metabolism (decrease in plasma cholesterol) and postprandial glucose metabolism (reduced plasma glucose levels). For example, daily consumption of 5 g of β-glucan has been found to significantly reduce serum total and LDL cholesterol in both hypercholesterolemic and healthy human subjects. Similar dietary levels (4–8 g/day) of β-glucan have also been shown to cause significant reductions in postprandial glucose and

insulin levels in diabetic and healthy human adults. The postulated mechanism for the physiological effects of β-glucan is related to its ability to form a hydrated viscous mass in the GIT. The increased viscosity of GIT contents leads to trapping and reduced absorption of glucose and bile acids, which reduces their plasma levels and enhances excretion in the feces. Reduced absorption of cholesterol also induces higher rate of cholesterol synthesis by the liver because of the need to produce more bile acids. Apart from the physical effects, fermentation of β-glucan in the colon produces large amounts of propionate, a short-chain fatty acid that inhibits cholesterol synthesis. The action of propionate is thought to be mediated through inhibition of activity of hepatic HMG-CoA reductase, a key hepatic enzyme involved in cholesterol synthesis. Because viscosity plays an essential role in the physiological benefits of β-glucan, processing methods that decrease size of the polymer could reduce potency of the compound. This is because it is well-known that for many polysaccharides, the length of the polymer chain is directly proportional to viscosity. Therefore, some of the inconsistencies observed in literature with regard to potency of β-glucan may be due to the use of polymers that differ in molecular size. Human intervention trials with low-molecular-weight β-glucans (80–370 kDa) showed no effects on serum lipid profiles whereas a higher-molecular-weight form (1,200 kDa) effectively reduced serum cholesterol levels. There are also differences in the solubility properties of β-glucans from different plant species or varieties from same crops. For example, barley β-glucan has substantially higher solubility than oats β-glucan.

1.5   Insoluble Fiber (IF)

These are polysaccharides such as cellulose and hemicelluloses that have reduced interactions with water and do not form the type of highly hydrated masses typical of soluble fibers. The IF is undigested in the upper gastrointestinal tract but is fermented by colonic microorganisms to form SCFAs that are known for their health benefits. IF increases the rate that nutrients or foods move through the gastrointestinal tract, which reduces amount of nutrients absorbed, especially glucose. Thus, dietary IF is known to have beneficial effects on blood glucose management in diabetic conditions. Dietary IF accelerates secretion of glucose-dependent insulinotropic peptide (GIP), an incretin hormone that stimulates postprandial release of insulin. Reduction in appetite and hence less caloric intake are also associated with consumption of IF. In addition to these effects, dietary IF boosts the SCFA content of the colon as a result of microbial fermentation. Physiologically relevant SCFAs are mainly acetate, propionate, and butyrate, all of which are present at approx. 80, 130, and 13 mM in the descending colon, cecum, and terminal ileum, respectively. Acetate serves as a substrate for hepatic de novo synthesis of lipids via acetyl-coA and fatty acid synthase (FAS). Propionate is known to suppress the lipogenesis-reduced expression of FAS, while butyrate serves as an important source of energy for colonic cells. SCFAs are readily absorbed into the colon, liver, and other tissues where they serve as sources of energy. It has been estimated that about 5–10% of the basal energy requirements of humans are provided by SCFAs. In addition to energy provision, SCFAs modulate various physiological processes such as secretion of satiety-inducing hormones (leptin, glucagon-like peptides, and peptide YY), immune or inflammatory responses, and cell proliferation/differentiation. By upregulating these hormones, high levels of SCFAs in the colon can contribute to feeling of fullness and lead to reduced food consumption with associated body weight benefits. It has been shown that SCFAs can inhibit free fatty acids (FFA) in the blood. This is important because FFAs inhibit glucose metabolism through inhibition of GLUT 4 transporters; therefore, reduction of FFAs by SCFAs reduces blood glucose levels. IF also may consist of polysaccharide-phenolic conjugates that are broken down in the colon to release the phenolic residues, mostly ferulic acid, diferulic acids, sinapic acid, p-coumaric acid, and caffeic acid. These phenolic compounds, especially ferulic acid, have potent antioxidant activities, and ferulic acid has been detected in human plasma following consumption of breakfast cereals. One of the health benefits associated with consumption of foods with high levels of IF is the sustained release of phenolic acids into the blood. In the colon, IF is acted upon by β-glucosidases and esterases produced by the gut microorganisms to release ferulic acid, which is then absorbed into the blood to reduce the risk of LDL and triglyceride oxidation. Thus, part of the health benefits of IF is due to fermentation of the polysaccharides to give SCFAs and release phenolic acids for absorption into the blood where they help to preserve the structural integrity of lipids. However, it has been noted that the IF in wheat and rice bran that is resistant to fermentation in the colon is not very susceptible to bacteria breakdown. But the dietary resistant IF from wheat and rice bran helps contribute to increased stool bulk and weight because of their water-holding capacity.

1.6   Resistant Starches (RS)

These are starch molecules that contribute fewer calories than regular starch molecules during digestion in the gastrointestinal tract. In essence RS molecules have lower glycemic index (ability to increase blood glucose level) when consumed as part of a normal diet. The use of RS in food and nutritional products could help people control the level of blood glucose and may be viewed as part of a diet to help prevent or reduce the impact of metabolic disorders such as obesity and type 2 diabetes. Consumption of RS at moderate doses has also been shown to increase moisture content and bulk of feces, which are important for eliminating toxic compounds and preventing physical injury to surface of the colon during bowel movement.

1.6.1 Definition

RS refers to a starch fraction and its degradation products that are resistant to enzyme digestion and will pass unchanged and unabsorbed from the stomach to the small intestine of healthy individuals. In the large intestine, the RS particles are fermented to variable extents by colonic microflora. About 30–70% of RS is fermented to form SCFAs by microorganisms in the colon, while the remaining portion is usually excreted in the feces. Processing treatments and individual differences contribute to the wide range of RS fermentation. First the RS is broken down into glucose by bacteria enzymes (α-amylases, glucoamylase, and isomaltase) followed by metabolic conversion of the glucose (via pyruvate) to form SCFAs (90% as acetate, propionate, butyrate) in addition to gases (CO_2, H_2, CH_4, etc.) and heat. SCFAs promote intestinal health because they are the preferred respiratory fuel for colon cells (colonocytes). Regular supply of SCFAs enhances blood flow within the colon, decreases luminal pH, and reduces the risk for development of abnormal colonic cell population. SCFAs also stimulate and enhance contraction of the colon muscles, which enhances degree of oxygenation and nutrient transport. Researchers have suggested that butyrate is the preferred substrate for colonocytes with data showing in vitro inhibition of the growth of transformed cells, tumor cell suppression, and decreased proliferation of colon mucosal cells. Butyrate also promotes maintenance of normal cell phenotype by increasing DNA stabilization and repair. There are three different types of RS:

(a) RS1 products are the physically inaccessible starches that are trapped in cellular matrices such as found in whole or partially milled seeds. If consumed as is, rate of digestion is slow, and there is only partial hydrolysis. If the seeds are properly milled into flour prior to consumption, the starch undergoes total digestion.

(b) RS2 products are the native uncooked starch granules such as found in raw potato or banana starches; they have a crystalline structure that reduces susceptibility to enzyme digestion. Rate of digestion is very slow, and there is only little hydrolysis in the gastrointestinal tract. When the starch is *freshly* cooked, total digestion takes place.

(c) RS3 products are found in cooked products as the retrograded starch portions formed at low or room temperatures. There are two main types. The first type is retrograded starch found in stale breads, cooked and cooled potato. Rate of digestion is slow, and the starch undergoes only partial hydrolysis. Because the retrograded chains are composed of amylose and amylopectin held mostly by the noncovalent hydrogen bonds, digestibility can be improved by reheating the starch to form a more loose structure that can be better accessed by digestive enzymes. The second type is retrograded amylose also found in breads and cooked and cooled potato. Rate of digestion is zero, and the starch is totally (100%) resistant to digestion. The amylose chains have undergone intense intermolecular bonding such that the digestibility cannot be improved by reheating and the resistance to enzyme hydrolysis is irreversible.

Resistance of starch to enzyme digestion in the gastrointestinal tract can be due to several factors, as discussed below:

1. Presence of enzyme-inhibiting nonstarch compounds in the food matrix. Tannic acid inhibits activities of amylase and maltase through protein precipitation and change in enzyme structure that destroys the active site. Other compounds such as lectins, hemagglutinins, polyphenols, and phytic acid that are present in legume seeds can reduce starch digestion.

2. The ratio of amylose/amylopectin varies among different crops and even cultivars. Amylose is a straight chain polymer, and this structural arrangement limits accessibility of the two terminal glucose units to β-amylase. In contrast, amylopectin is highly branched which makes several terminal end glucose units accessible to β-amylase for digestion. Also during cooling of cooked foods, amylose forms indigestible retrograded units faster than amylopectin. Therefore, RS character is directly proportional to the level of amylose in the food product.

3. Presence of amylose-lipid complexes could reduce accessibility of glycosidic bonds to enzyme digestion. However, this effect could be overcome by increased levels of amylase.

4. Particle size of the food determines surface area and rate of starch digestion. In food products with smaller particles, there is a larger surface area, which favors rapid digestion of starch when compared to large particles that have small surface areas.

5. RS levels in raw foods such as potatoes, beans, lentils, and bananas can range from 14% to 47% depending on source and other environmental conditions. In processed foods, RS levels can be as high as only 5% as found in cooked beans, because in some cases, the RS1 and RS2 in raw foods are destroyed during processing. However, processing factors such as freeze drying, cooking, autoclaving, and extrusion can lead to production of RS3 products, which is mostly associated with retrogradation of the linear chains of amylose. Other factors that affect amylose retrogradation include temperature, water content, pH, number of heating or cooling cycles, incubation time, polymer chain length, and presence or absence of sugars and lipids.

1.6.2 RS and Blood Lipids

Several animal experiments have demonstrated the ability of RS to reduce fat accretion, as well as lower blood cholesterol and triglycerides,

though the results need to be confirmed in human subjects. However, in humans, the consumption of RS led to accelerated intestinal transit rate, which reduced absorption of bile acids. RS consumption may also reduce concentrations of cholesterol and triglycerides in the liver. In the case of the RS2 high-amylose cornstarch or raw potato starch, the hypocholesterolemic effect seems to be mostly through increased fecal excretion of bile acids but not fermentation. This is because RS can bind bile acids or cholesterol in the intestine and prevent reabsorption into the blood (through the portal vein). Reduced availability of cholesterol through the portal vein induces the liver to remove more LDL cholesterol from the circulatory system, which results in reduced blood cholesterol levels. Dietary RS has been found to reduce plasma cholesterol and triglyceride levels by as much as 32% and 42%, respectively. In fact RS was found to be more effective in reducing blood lipids than the drug cholestyramine that is used as a bile sequestrant.

1.6.3 RS and Enhanced Mineral Absorption

Cecal hypertrophy accompanies the acidic fermentation that occurs following consumption of RS, which then enhances calcium uptake in the colon. This is because of an increased surface area that accompanies cecal hypertrophy and also contributions from the SCFA. Production of SCFA could enhance calcium absorption through modification of electrolyte exchanges such as Ca-H; such a less charged calcium ion can be readily absorbed through passive diffusion through the cell membrane. RS consumption also increases magnesium absorption because of increase ion solubility in the acidic environment that results from fermentation. Because magnesium solubility is generally higher than that of calcium, the combined effects of cecal hypertrophy and acidic fermentation lead to enhanced magnesium absorption. Absorption of iron, zinc, and copper was also significantly increased in RS-fat rats when compared to the control group.

However, it is important to note that the distribution of SCFA along the colon of rats is different from that of humans, and direct extrapolation of results from rats to humans may not be accurate.

1.6.4 RS and Control of Blood Glucose

In order to maintain regular blood glucose levels and control the degree of postprandial hyperglycemia, a balance between glucose influx (from dietary sources) and glucose removal (insulin dependent) must be maintained. Therefore, dietary RS can help control blood glucose because it reduces glucose influx due to reduced rate of digestion and amount of absorbable glucose in the upper gastrointestinal tract. The slow release of glucose by RS also has beneficial effects on insulinemia by reducing insulin response. Thus, RS starch consumption could help control blood glucose in clinical conditions such as diabetes and impaired glucose tolerance. In addition, the limited availability of glucose encourages the body to use stored fat for energy and reduce mass of adipose tissue, which could help maintain healthy weight.

1.6.5 RS and Risk of Developing Colon Cancer

Very few epidemiological studies exist for the relationships between RS in the diet and risk of colon cancer. But in general, populations that have high dietary intake of RS have lower risk of colon cancer. This is supported by the fact that biopsy studies have shown a strong inverse relationship between butyrate levels and aberrant crypt proliferation in the rectum. It has also been suggested that RS lower colon cancer risk because of their ability to reduce concentration of secondary bile acids and rate of colonic mucosal cell proliferation. Rats fed about 6% retrograded high-amylose cornstarch in the diet had reduced number of dimethylhydrazine- and azoxymethane-induced aberrant crypt foci, suggesting that RS may be effective in the early stages of colon cancer. In various animal experiments, dietary intervention with RS in the later stages of colon cancer was found to be ineffective in preventing tumor growth.

1.7 Slowly Digestible Starch (SDS)

This is the intermediate starch category between rapidly digestible starch (RDS) and RS. Typically, SDS consists of an optimal mixture of amorphous and semicrystalline carbohydrate polymers. High content of amylopectin or high degree of branching within the starch granular structure is also favorable characteristic for SDS. SDS molecules have a slow and sustained effect on postprandial blood glucose levels and therefore have moderate impact on glycemic index (GI). Diets that exhibit low GI are associated with reduced risk of diabetes and cardiovascular diseases, while high GI foods can be a risk factor for colon and breast cancer. In contrast to RDS that is completed digested within 20 min, digestion of SDS takes place over a range of 20–120 min. In general, RDS is mostly amorphous in structure, which enhances access by digestive enzymes, whereas SDS is composed of both amorphous and crystalline structures. The presence of a crystalline structure reduces accessibility of the SDS to digestive enzymes, thereby increasing digestion time. Native cornstarch is an example of SDS, though there is currently no commercial form of SDS.

Physiological effects of SDS include improved metabolic profile especially lower postprandial insulinemia and lower levels of blood triglycerides. SDS can be used as a dietary factor to reduce the risk of developing metabolic syndrome, which is associated with insulin resistance and cardiovascular disorders. SDS has been shown in obese, insulin-resistant patients to improve metabolic profile such as reduced levels of circulating lipids (triglycerides, triglyceride-rich lipoproteins, and apolipoproteins) and lower postprandial insulinemia. SDS is also a suitable food ingredient for the reduction in meal-associated hyperglycemia in diabetic patients. Consumption of SDS can improve carbohydrate metabolism and reduce insulin requirement of type 2 diabetic patients. For example, consumption of SDS increased secretion of gut incretin hormones, glucagon-like peptide-1 (GLP-1), and glucose-dependent insulinotropic polypeptide (GIP) during the late postprandial phase (180–300 min postconsumption). The delayed increases in GLP-1 and GIP could help maintain normal glucose homeostasis and energy storage, which is of benefit to patients with disorders (e.g., diabetes) involving glucose metabolism. Consumption of SDS-containing foods as part of a breakfast diet is helpful in improving carbohydrate metabolism and reduction of insulin requirements in type 2 diabetes patients that use insulin to manage their blood glucose level. Due to the lack of a commercial SDS product, native uncooked cornstarch is recommended for the management of insulin-treated type 2 diabetes. SDS may also be good for increased satiety and reduced need to eat because of the ability to lower insulin response following meal consumption.

1.8   Prebiotics

1.8.1   Definition

Dietary carbohydrates like RS, IF, and soluble fiber that are able to stimulate, specifically the growth of potentially beneficial bacteria, e.g., bifidobacteria at the expense of the more harmful pathogenic microorganisms, are called prebiotics. Presence of prebiotics in the colon helps to modify the microflora in such a way that the health-promoting bacteria like bifidobacteria and lactobacilli become predominant in numbers and may be accompanied by elimination of pathogenic bacteria. In essence, consumption of prebiotics enhances the development of a healthy gut. Prebiotics are not digested or absorbed in the stomach and small intestine but are fermented once they reach the large intestine. Therefore, they are termed "colonic food," which refers to a food that enters the colon to become a substrate for the endogenous bacteria and indirectly provides the host with energy and metabolic substrates. In general, compounds can be classified

as prebiotics based on three main criteria: (1) resistance to digestion and absorption in the upper tract (stomach and small intestine) of the gastrointestinal tract, (2) susceptibility to microbial-induced fermentation in the lower intestine (colon), and (3) selective stimulation of the growth and activity of intestinal microorganisms that have positive effects on human health. Potential health benefits of prebiotics include increased bioavailability of minerals, and reduced risks of various diseases such as cancer, intestinal infections, cardiovascular disorders, obesity, and diabetes. Indigestible carbohydrates such as oligofructose and inulin are the most studied forms of prebiotics because they preferentially stimulate the growth of a health-promoting bacteria population that is dominated by bifidobacteria. Some of the potential health benefits associated with consumption of prebiotics include:

(a) Normalization of stool frequency and consistency. Normal water content of stool is between 70% and 80%, and consumption of prebiotics has been shown to reduce the incidence of very loose or very hard stools. Frequency of stool output was also increased by prebiotics from average of 1.1 to up to 6.7 times per week in pregnant women who suffer from constipation. It should be noted that high levels of prebiotics in the diet could cause increased incidence of flatulence due to high levels of gases associated with microbial fermentation.

(b) Antioxidant effects. Some prebiotics such as arabinoxylans (ABX) and arabinoxylan-oligosaccharides (ABXO) contain covalently bound hydroxycinnamic acids of which ferulic acid is the most abundant. ABX (Fig. 1.4) are polysaccharides present mostly in the bran of cereal grains, especially wheat, barley, and rye. ABX polymers are made up of a xylan backbone with L-arabinofuranose (5-atom ring form of L-arabinose) attached randomly through $1\alpha \rightarrow 2$ and/or $1\alpha \rightarrow 3$ linkages to the xylose units throughout the chain. ABX is classified as a pentosan because xylose and arabinose are pentose sugars. ABXO is usually the product obtained after hydrolysis of ABX by microbial xylanases in the colon.

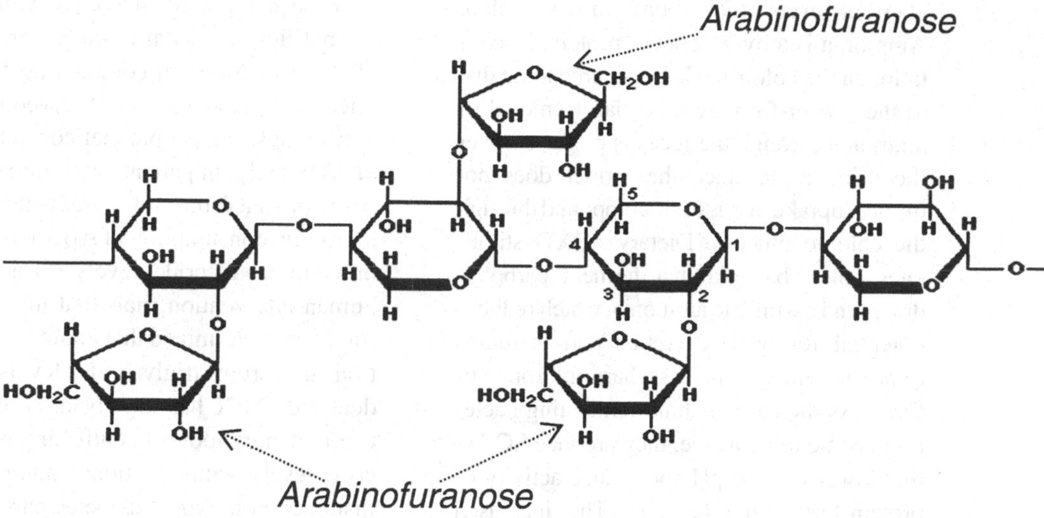

Fig. 1.4 Typical repeating unit of arabinoxylan

However, ABXO may also be generated in cereal-based foods (bread, cookies, beer, pasta) through the action of endogenous xylanases on ABX or xylanases produced by contaminating microorganisms. Beer has been shown to be a very rich source of ABXO. Ferulic acid has exhibited potent in vitro antioxidant properties with evidence suggesting potential anticancer efficacy especially against breast tumors. Evidence also shows that the ABXO-ferulic acid complex could have higher antioxidant potency than ferulic acid alone with strong inhibition of copper-induced oxidation of low-density lipoproteins. The ABXO-ferulic acid is a unique antioxidant complex that remains undigested and unabsorbed until in the colon where it is broken down into ABXO and free ferulic acid by bacterial feruloyl esterases. In diabetic rats, treatment with ABXO led to substantial reduction in level of serum lipid peroxidation, which could be beneficial toward reducing the risk of atherosclerosis and coronary heart disease.

(c) Immune-stimulating effects. Inulin and other fructooligosaccharides (FOS) have been shown to stimulate activity of natural killer (NK) cells and enhance SCFA-mediated anti-inflammation responses. Treatment of lectin-stimulated mice spleen cells with corn-bran-derived ABX led to significant increases in interleukin-2 (IL-2) and interferon-γ, an indication of immune stimulation. In rats, inclusion of ABX in the diet led to development of smaller sized colon tumors accompanied by increased activity of spleen NK cells. Allergic response in atopic dermatitis was attenuated when corn bran ABX was included in the mice diet, which suggests anti-inflammatory properties.

(d) Anticarcinogenic effects. Under normal conditions, glucuronic acid conjugates are used by the body to inactivate toxic compounds or xenobiotics and render them difficult to absorb but easy to eliminate through the urine of feces. However, activity of β-glucuronidase leads to hydrolysis of the glucuronic acid conjugates and release of toxic compounds, especially into the colon. In a human intervention study, it was shown that dietary ABX enhanced reduction in the level of β-glucuronidase, which reduces the potential for toxin-induced formation of cancer cells. This was demonstrated in chemical carcinogen-treated rats where addition of xylooligosaccharides (XOS) and FOS to the diet led to reduction in the number of aberrant crypt foci found in the colon. Ammonia is a by-product of protein fermentation in the colon and is considered toxic due to the cancer-forming potential. Removal of ammonia through the feces is preferred over the urine route since the former does not involve uptake of this toxic compound through the colonic mucosa. Dietary ABXO stimulates colon bacteria that ferment carbohydrates and assimilate ammonia, which reduces potential for urine excretion and increases removal through the less harmful route of feces. As the carbohydrate-fermenting bacteria become more active, they produce SCFAs that lower colonic pH and reduce activity of protein-fermenting bacteria. The increased acidity (higher levels of H^+) resulting from activity of the carbohydrate-fermenting bacteria enhances protonation of ammonia to form the charged ammonium ion that is less easily absorbed through the colonic mucosa due to the reduced ability of charged groups to pass through the cell membrane lipid bilayer. Thus, the ammonium ion is removed mostly through the feces and less through the urine. For example, dietary XOS was found to reduce serum ammonia level in patients that suffer from liver cirrhosis, suggesting increased elimination through the feces as opposed to the urine.

(e) Antimetabolic syndrome effects. Prebiotics such as XOS and ABXO have been shown to reduce the risk of cardiovascular diseases through reductions in blood levels of triglycerides and cholesterol as was demonstrated in rats treated with a diabetes-inducing drug. XOS also attenuated increases in cholesterol, triglycerides, and total body, liver, and abdominal fat that are normally associated with consumption of high cholesterol and high-fat diets in experimental rats. On the other hand, dietary ABX only reduced the levels of cholesterol in serum and liver, but level of triglycerides was unaffected. Results from human intervention trials vary and have not been consistent. For example, slight reductions in serum cholesterol and triglycerides

have been reported in young women with normal lipid levels after daily consumption of 2.7 g of XOS. In contrast, no beneficial effect on lipid levels was observed in elderly normolipidemic people that consumed 3.8 g of XOS daily. In patients with impaired glucose tolerance but not in diabetic patients, long-term consumption of ABX led to reductions in triglyceride levels. From various human intervention trials that involved inulin, it was determined that about 7.5% reduction in serum triglyceride levels can be detected. ABX has also been shown to be useful dietary tool for controlling blood glucose levels with additional attenuation of insulinemia in type 2 diabetes patients, people with impaired glucose tolerance, and even healthy individuals. Supplementation of rat diet with ABXO or XOS had beneficial effects in diabetic rats through reductions in body weight and attenuation of the increase in blood glucose levels. The beneficial effects of prebiotics on lipid and glucose metabolism may be due to increased levels of colonic fermentation products such as SCFAs that are taken up into the liver where they inhibit lipogenesis. The SCFAs may also work by stimulating intestinal production of glucagon-like peptide 1 (GLP-1). GLP-1 stimulates insulin synthesis, while increased level of insulin will reduce lipolysis but increase glycogen synthesis. In addition, ABX has a higher degree of polymerization and hence higher viscosity than ABXO and XOS. Thus, similar to observed effects of other viscous fibers, ABX could impact beneficial effects on human health through delay of gastric emptying (increased satiety) and reduction in the rate at which digested nutrients diffuse to the absorption interface. The reduced nutrient diffusion rate enhances trapping of cholesterol within the digesta and reduces uptake of dietary cholesterol into the blood circulatory system.

Typical examples of prebiotic carbohydrates include inulin, oligofructose, and lactulose.

Table 1.1 Inulin content of some foods

Source	Inulin content (%)
Asparagus	2–3
Chicory (white carrot)	15–20
Dandelion greens	12–15
Garlic	9–11
Jerusalem artichoke	16–20
Leek	3–10
Raw onion	2–6
Dried onion	5–32
Wheat	1–6

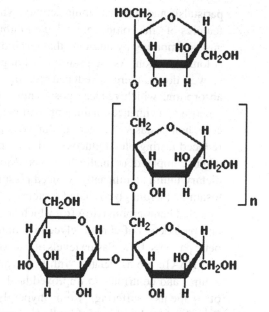

Fig. 1.5 Chemical structure of inulin showing linear fructose chain with glucose at the head (n is ~35)

1.8.2 Inulin

This prebiotic, also known as raftiline, is found in several foods such as wheat, onion, garlic, bananas, fruits, and vegetables, but industrial production utilizes chicory (white carrot) roots (Table 1.1). Inulin is extracted from chicory roots through hot water extraction followed by refining and spray drying into a powder that is composed of a mixture of linear molecules with a basic G-Fn chemical structure. G represents a glucosyl moiety, while Fn are the fructose moieties joined together by β-2,1 glycosidic bonds, with n (degree of polymerization, DP) ranging from 3 to 60 with an average value of 10. The repeating unit of inulin chain is shown in Fig. 1.5. Some long-chain inulin products with a DP of about 25 are obtained after lower DP fractions have been physically removed from the initial mixture. Inulin is partially soluble in water where it forms slightly thickened product that has a bland neutral taste without any off-flavor and can contribute to body and mouthfeel of food products. Because of the gelling ability, inulin has the capacity to be used as a fat replacer with excellent mouthfeel characteristics and can act as emulsion and foam stabilizers.

1.8.3 Oligofructose

This prebiotic also known as raftilose is obtained through enzymatic hydrolysis of inulin to give a product composed of linear G-Fn and Fn chains with DP ranging from 2 to 8 (average of 4). Therefore, unlike inulin chains that contain terminal glucose units (G-Fn), oligofructose can contain linear fructose chains that are devoid of glucose (Fn) in addition to the G-Fn chains. Due to the shorter chain length and higher contents of fructose (Fn chains), oligofructose is much more soluble and moderately sweet (with clean taste) when compared to inulin. When combined with high-intensity sweeteners, oligofructose produces a rounder mouthfeel and a more sustained fruity flavor with little aftertaste. A combination of inulin and oligofructose provides better organoleptic properties, and they are already being used in food products such as yogurts, dairy drinks, dessert, and meal replacers.

1.8.4 Inulin, Fructooligosaccharides, and Oligofructose as Bioactive Prebiotic Compounds

In general, inulin consumption has been reported to delay gastric emptying time, which provides a longer satiety effect and reduces frequency or quantity of food intake. This effect is attributed to the production of SCFAs in the colon during inulin fermentation. SCFAs are believed to inhibit peristalsis and stimulate tonic activity, which are features of "ileocolonic brake," the inhibition of gastric emptying by nutrients that reach the ileocolonic junction. As the gastric emptying rate is slowed down, there is reduced rate of nutrient absorption, which reduces postprandial plasma glucose and enhances maintenance of better glucose homeostasis. In fact, inulin consumption reduced fasting blood glucose and insulin levels when compared to inulin-free diet. Moreover, dietary inulin significantly reduced plasma fructosamine (serum glycosylated protein) as well as glycated hemoglobin (HbA1c), which are important parameters of chronic glycemic control diabetic patients. Low plasma levels of fructosamine and HbA1c are associated with better glycemic control and contribute to improved health status of patients suffering from hyperglycemia. Beneficial health effects of inulin are also due to ability to reduce gastric emptying time through increased stimulation of the secretion of gut-modulating peptides such as neurotensin (NT) and somatostatin (SS). NT and SS are peptide hormones that influence gastroduodenal motility by reducing gastric emptying rates. By reducing gut motility, inulin attenuates rise in blood sugar level associated with meal consumption, which is helpful for diabetic patients in preventing hyperglycemia. Supplementation of diet of elderly subjects with fructooligosaccharides resulted in increased fecal output of *Bifidobacteria* along with decreases in plasma cholesterol and lipid peroxides. The increase in *Bifidobacteria* population leads to greater assimilation and catabolism of cholesterol, which is reflected in the reduced plasma cholesterol level that is associated with increased consumption of fructooligosaccharides. The increased growth of beneficial bacteria population leads to high rate of glycolytic activity in the colon, which enhances conversion of inactive dietary glycosides into aglycones that are more bioactive and readily absorbed. Data also suggests that prebiotic carbohydrates have antioxidant effects as evident in the reduced lipid peroxide production. There is also an increase in colon tissue mass because of the enhanced proliferation of healthy colonocytes due to abundance of energy in the form of SCFAs.

1.8.4.1 Effect on Colonic Bacteria Population

Human beings have no digestive enzyme that can hydrolyze the β,2-1 bonds in inulin and oligofructose, and so they remain structurally intact during passage through the upper gut. Because they pass intact into the colon, dietary inulin and oligofructose are quantitatively fermented by colonic microflora where they selectively promote the growth and multiplication of bifidobacteria. Daily consumption of a few grams of inulin or oligofructose by human subjects has been shown to modify fecal flora such that the bifidobacteria become the dominant genus accompanied by substantial reductions in population of harmful bacteria such as the clostridia.

1.8.4.2 Inulin and Oligofructose as Dietary Fiber

Regular consumption of inulin and oligofructose leads to increases in colonic biomass, stool weight, and frequency. Microbial fermentation converts inulin and oligofructose into short-chain fatty acids (SCFAs, mostly acetate, propionate, and butyrate) and lactate, which leads to reduction in colonic pH. Low pH favors the growth of beneficial bacteria like the lactic acid and bifidobacteria types but is detrimental to the growth and survival of pathogenic bacteria, especially the nonacidophilus types. Moreover, low pH conditions increase protonation of harmful amines (potential carcinogens), which enhances expulsion with the feces since the protonated amines are less absorbed than the neutral amines. This is because the charged (protonated) amines are less able to diffuse through the lipid bilayer of cell membranes when compared to the neutral forms. The presence of high levels of amines can stimulate carcinogenesis and progress into tumor development. In experimental rats, inulin reduced the incidence of aberrant crypt foci (ACF) that is induced by colon carcinogens such as azoxymethane or dimethylhydrazine; ACF is recognized as a marker for risk of colon cancer formation. The mechanism is essentially related to ability of fermentation products from inulin, especially the SCFAs, to inhibit bacteria-mediated conversion of primary (cholic and chenodeoxycholic) to secondary (deoxycholic and lithocholic) bile acids. This is because secondary bile acids have cocarcinogenic and tumor-promoting properties. For example, there is a negative correlation between amount of butyrate and bile acid metabolism in the colon. Apart from the antitumor effects, consumption of inulin has been shown to reduce serum levels of triglycerides and cholesterol, mostly through decrease in plasma concentrations of very-low-density lipoproteins (VLDL). Decreased concentration of VLDL was due to decreased hepatic synthesis of triglycerides arising from reductions in the activities of lipogenic enzymes. Inulin consumption also increases plasma level of HDL-cholesterol, which has inverse relationship with formation and growth of atherogenic plaques.

1.8.4.3 Relationships with Calcium and Osteoporosis

Osteoporosis is associated with development of soft bones due to reduced plasma levels of calcium and increased bone resorption (loss of calcium from the bones to compensate for low plasma levels). Postmenopausal women are especially susceptible to osteoporosis due to reduced estrogen production because the hormone is believed to improve calcium utilization and bone development. Apart from increased calcium intake, a more efficient intestinal absorption is an alternative way of boosting plasma calcium levels and preventing bone resorption. Consumption of oligofructose-enriched inulin promotes improved dietary calcium absorption and could be used in the formulation of functional foods for the prevention of osteoporosis in postmenopausal women. Similarly, oligofructose-enriched inulin has been found to increase bone mineral density in adolescents. A potential mechanism for the increased calcium absorption activity of oligofructose is the higher solubility when compared to inulin. Solubility of the prebiotic-calcium complex enhances hydrolytic activity in the gut, which frees up calcium for absorption. This is because it is well-known that reduced solubility of inorganic sources of calcium is a contributory factor for their poor calcium bioavailability properties when compared to the more soluble organic calcium complexes.

1.8.4.4 Relationships with Iron Absorption

Conversion of inulin into short-chain fatty acids by beneficial bacteria leads to a reduction in pH of the colon, which enhances solubility of iron. Reduction in pH also increases proliferation of mucosa cells, which provides increased surfaced area for iron absorption. Within the mucosa cells, it is believed that metabolic products from inulin fermentation can activate the gene named "divalent metal transporter 1," which encodes for an iron transport protein. The overall results of increased iron solubility, enhanced absorption, and transportation are the increased bioavailability of dietary iron. Therefore, inulin consumption may provide therapeutic relief for sufferers of iron deficiency and help boost blood hemoglobin levels.

1.8.4.5 Mechanisms Involved in Prebiotic-Enhanced Mineral Absorption

Increased solubility of minerals as a result of increased bacteria production of SCFA, enlargement of the absorption surface by promoting enterocyte proliferation (mediated by bacterial fermentation products such as lactate and butyrate), and increased expression of calcium-binding proteins are some of the mechanisms that have been proposed to explain the enhanced mineral absorption associated with dietary prebiotics. The following proposed mechanisms are based on results from a variety of animal, cell culture, and human intervention studies:

- Improved gut health and stimulation of immune defense: oral administration of oligofructose-inulin mixture increases number of goblet cells as well as the thickness and composition of the colonic epithelial mucus layer. Mucins became more acidic because of increased production of sulfomucins, which indicates a more stabilized mucosa, which contributes to better gut health because of improved absorption functions. As the mucosa flora becomes more stabilized, there are reduced gastrointestinal infections and oxidative damage to enterocytes. In experimentally induced lipid peroxidation in enterocyte liposomes, the lactic acid bacteria strain, *Streptococcus thermophilus* YIT 2001, effectively inhibited mucosa damage. There is also decreased formation of proinflammatory cytokines, especially TNF-α during acute diarrhea, which could reduce degree of enterocyte inflammation.

- Production of bone-modulating factors: consumption of fructooligosaccharides (FOS) can increase the bone-preserving effects of phytoestrogens as a result of increased isoflavone bioavailability. It is well-known that flavonoids such as isoflavones are more efficiently absorbed from the intestinal lumen when presented in the aglycone rather than the intact glycoside. Intestinal FOS stimulates growth of bifidobacteria and lactobacilli that produce β-glycosidase, an enzyme that converts the poorly absorbed glycosides into the better absorbed free aglycones through hydrolysis of the glycosidic bonds in flavonoids. β-glycosidase also converts daidzein (an isoflavone) to equol (a nonsteroidal estrogen), the isoform with a higher bone-preserving potential.

- Intestinal flora stabilization: FOS can prevent the increase in cecal pH that accompanies antibiotic treatments. During antibiotic treatment, there is reduced production of SCFA (hence increased cecal pH) as a result of losses in the beneficial microbial population, which causes reductions in mineral solubility, absorption, and deposition into the bones. Dietary FOS prevents the antibiotic-mediated loss in beneficial microbial population and stabilizes the ability of the colon to maintain regular levels of SCFA that will promote mineral solubility and absorption.

It is possible that the beneficial effects of prebiotics may be dependent on the characteristics of the individual such as:

- Physiological age: children who are just reaching the age of puberty or adults in postmenopausal years will have a high requirement for calcium and will benefit more from consumption of prebiotics than healthy adults.

- Stage of postmenopause: bone loss in women who are in the very early years of menopause is more affected by the decline in circulating estrogen when compared to women in the later years. Therefore, women who are more than 6 years postmenopause may benefit more from increased levels of dietary prebiotics than women in the very early stages.

- Capacity to absorb calcium: dietary supplementation with prebiotics (and calcium) will benefit people with low calcium intakes and upregulated absorption capacity but not people with filled calcium stores and low concentrations of vitamin D3 or low mucosal expression of calcium-binding proteins.

Summary of metabolic fate of prebiotics
- Inulin and oligofructose are not hydrolyzed in the mouth, stomach, and small intestine.
- In the large intestine, they undergo complete anaerobic fermentation by bacteria.
- They do not contribute any calories.
- They are completely fermented in the colon, so inulin is not excreted in the stool.

Summary of potential health benefits of prebiotics
- Increase bioavailability of minerals such as calcium, magnesium, and iron
- Reduce the risk of colon cancer
- Reduction in cholesterol and blood lipids
- Prevention of gastrointestinal tract infections
- Increased growth of bifidobacteria, which has the following beneficial effects:
  (a) Produces nutrients such as B-group vitamins and folic acid.
  (b) Produces digestive enzymes.
  (c) Reduces food intolerance by utilizing residual nutrients from the upper gut.
  (d) Improves nutrient management.
  (e) Reduces liver toxins, i.e., blood amines and ammonia, by using them as fuels.
  (f) Competitive elimination of pathogenic microorganisms. This is believed to be due to the fact that only the bifidobacteria are capable of digesting or fermenting the prebiotic carbohydrates because they produce β-fructosidases. Therefore, abundance of dietary prebiotics allows the bifidobacteria to grow at the expense of the harmful bacteria.

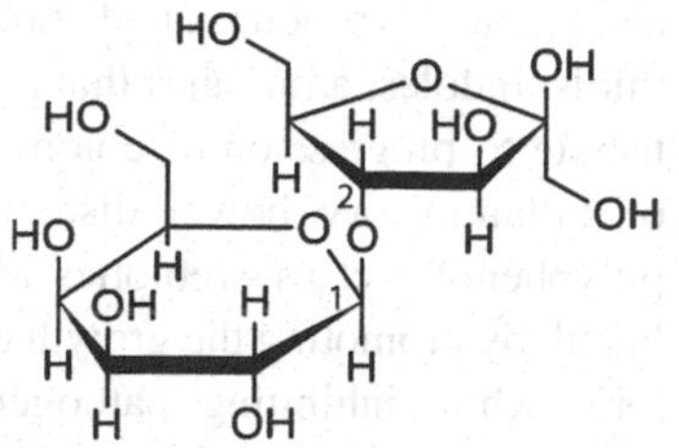

Fig. 1.6 Structural configuration of lactulose showing the C1→C2 bonding

1.8.5 Lactulose as Prebiotics

Lactulose (Fig. 1.6) is a synthetic disaccharide analog of lactose consisting of fructose chemically bonded to galactose. Lactulose does not occur naturally but is produced by isomerization of lactose in basic media and in the presence of different types of catalyst. However, lactulose can be formed during heat processing (especially ultrahigh temperature) of raw milk due to the presence of bicarbonates and other minerals that provide a slightly basic environment. Due to the β-type of glycosidic bonding between fructose and galactose, lactulose (4-O-β-D-galactopyranosyl-D-fructose) is resistant to digestion in the upper part of the digestive tract. Since it is also not absorbed, lactulose passes intact into the large intestine where it is selectively fermented by *Bifidobacteria* and *Lactobacilli* to form gases and short-chain fatty acids. Lactulose fermentation leads to increase in fecal biomass and reduction of pH in the colon lumen, which facilitates conversion of toxic ammonia (NH_3) to the less toxic and easily excreted ammonium ion (NH_4^+). Because of its selective fermentation by *Bifidobacteria* and *Lactobacilli*, lactulose is added to infant formula to stimulate establishment of *Bifidobacteria* flora similar to that of breast-fed babies. Similarly, lactulose has been shown to be effective as a prebiotic because daily consumption of 10 g led to significant increase in *Bifidobacteria* population while decreasing *Clostridia* population. Dietary lactulose has been shown to promote mineral bioavailability in animal and human experimental trials. For example, rats fed lactulose-containing diets had increased absorption of calcium up to 10% level of the sugar in the diet. A mixture of fructose and galactose did not have the same effect, which confirms that it is the indigestibility of lactulose that contributes to the observed effect on calcium absorption and not the identity of the component monosaccharides. More importantly, lactulose must be present in the diet to stimulate calcium absorption since administration of lactulose alone after a 48-h period or up to 7 days prior to feeding of calcium diet had no effect. Removal of the rat colon did not negate the beneficial effects of lactulose, which indicates that the small intestine is the site of enhanced calcium absorption. In healthy human adults, absorption of calcium and magnesium was enhanced when coadministered with lactulose. A high concentration of digestion-resistant lactulose in the small intestine increases the amount of fluid within the lumen in order to maintain isotonicity. The presence of additional fluid increases distention and permeability of the intracellular junctions between enterocytes, which enhances passive absorption of calcium and possibly other ions. Lactulose also acts as a laxative and can be used to treat childhood constipation by reducing intestinal transit time, which increases number of fecal productions. Lactulose is also well established as a useful compound for the treatment of systemic encephalopathy, though actual mechanism is not fully understood. Addition of lactulose to soy milk increases the growth of *B. longum* and *B. animalis*, which convert inactive isoflavone glycosides to the biologically active isoflavone aglycones with reported estrogenic activities. This transformation has been shown to be increased when lactulose is added to probiotic cultures, especially during fermentation of soy milk with *Lactobacilli*. Thus, the beneficial effects of soybean isoflavones could be enhanced by increased consumption of lactulose in the diet.

1.9 Polyphenols as Prebiotics

Dietary flavanols are metabolized by phase I and II enzymes to form O-methylated, sulfated, and glucuronidated conjugates during absorption from the intestinal lumen into the portal vein. However, not all the flavanols are absorbed, and substantial amounts of compounds such as (-) epicatechin, (+)-catechin, and procyanidins reach the colon undigested. In the colon, these flavanols serve as nutrients for several beneficial microorganisms that catabolize the polymeric phenolics to form low-molecular-weight compounds such as phenylacetic, phenylpropionic, and phenylvaleric acids. The intestinal microflora also metabolizes epigallocatechin and epicatechin into 5-(3′,4′,5′-trihydroxyphenyl)-γ-valerolactone and 5-(3′,4′-dihydroxyphenyl)-γ-valerolactone. Importantly, the metabolism of these flavanols occurs in the presence of fructooligosaccharides, which suggests that the presence of other energy sources does not prevent utilization of these phenolic compounds by intestinal microflora. (+)-Catechin has been shown to have greater effects than (-)-epicatechin in modifying bacteria growth. Addition of (+)-catechin to a bacteria batch culture significantly reduced the growth of *C. histolyticum* group, but there were increases in the growth of beneficial groups of *C. coccoides-Eubacterium rectale*, *Bifidobacterium*, and *Lactobacillus*. *C. coccoides-Eubacterium rectale* possess the ability to generate SCFAs by saccharolytic metabolism, and these microorganisms are associated with beneficial effects at the cellular and systemic levels. This is because increased production of SCFAs benefits gut health because of their ability to inhibit proliferation of preneoplastic cells and accelerate conversion of cholesterol into bile acids. It is also known that *Lactobacillus plantarum* can degrade complex esters of gallic acid and glucose. Tannic acid is hydrolyzed to glucose and gallic acid, which is then decarboxylated to form pyrogallol. Hydrocinnamic acid is also metabolized through reduction of carbon-carbon double bond to produce hydroxyphenylpropionic acids, which is further decarboxylated to *p*-ethylphenols. The capacity of these probiotic microorganisms to utilize polyphenolic compounds as energy sources or growth substrates is an important aspect in promoting a healthy gut. This is because the enhanced growth of *Bifidobacterium* and *Lactobacillus* will increase production of organic acids (lactic and acetic) in the colon, which inhibits ability of pathogenic microorganisms to grow and colonize the gut epithelial lair. The ability of flavonols to decrease growth of *C. histolyticum* group of bacteria reduces proteolytic metabolism

and attenuates production of compounds (amines, thiols, indoles, ammonia) that are known to contribute to progression of colon cancer and onset of inflammatory bowel disease. Thus, dietary polyphenols act as prebiotics and modulate gut health by promoting the growth of beneficial bacteria while inhibiting pathogenic microorganisms. The prebiotic effects of polyphenols could be very important since most of the free polyphenols eventually reach the colon while some types of dietary fiber contain bound phenolics that are subsequently utilized in the colon. While these significant modulations of growth of health-promoting bacteria by flavanols have been demonstrated using in vitro bacteria culture works, future human studies are required to confirm the results.

1.10 Role of SCFAs in Inflammation

One of the key aspects of the beneficial effects of dietary fibers, prebiotics, and indigestible polyphenolic compounds is their ability to induce production of large quantities of SCFAs in the colon. Inflammation has been shown to be associated with various pathogenic conditions such as diarrhea, atherosclerosis, and renal dysfunction. SCFAs are able to mitigate the damaging effects of inflammation and may be viewed as an important bioactive agent through which health benefits of indigestible nutrients are manifested. To combat inflammation, SCFAs modulate various cell signaling pathways and production of eicosanoids and cytokines as well as attenuate enzyme activities.

In response to inflammatory agents, the body recruits several anti-inflammatory cells and produces anti-inflammatory compounds in order to reduce damaging effects associated with inflammation. During inflammation, leukocytes (white blood cells) will migrate from the bloodstream to the inflammation site after a series of steps that involve activation of proteins such as adhesion molecules (integrins, macrophage antigen-1, and selectins) and chemokines that interact with endothelial cells. SCFAs have G-protein-coupled receptor (GPCR)-dependent chemotaxis (directional migration) effects on leukocytes, which enhance the anti-inflammatory capacity of the body. This is because binding of SCFAs to GPCRs leads to activation of cell signaling pathways, e.g., MAPK, protein kinase C, and activating transcription factor-2 (ATF-2), that produce various chemoattractants and molecules involved in leukocyte recruitment. For example, genetic alterations and use of drugs that inhibit these signaling pathways have been shown to substantially reduce GPCR-dependent leukocyte recruitment during inflammation. SCFAs also have direct effects as anti-inflammatory agents because they modulate production of monocyte chemoattractant protein-1 (MCP-1), a compound that attracts macrophages and lymphocytes. The ability of MCP-1 to recruit and accumulate macrophages in adipose tissue is an important causative factor for development of insulin resistance during obesity. Macrophages also produce nitric oxide and cytokines (TNF-α, IL-6, IL-1β, thromboxane A$_2$, prostaglandins E$_2$ and F$_{1\alpha}$) that are involved in other chronic diseases such as Alzheimer's, atherosclerosis, and rheumatoid arthritis. SCFAs, especially butyrate, can reduce MCP-1 production by adipose tissue and may be responsible for some of the beneficial effects on insulin resistance and glucose metabolism that are associated with increased consumption of prebiotics. SCFAs also modulate the inflammatory mediator activity of macrophages through attenuation of histone deacetylase (HDAC) activity. It has been shown that butyrate and acetate are most and least potent inhibitors of HDAC, respectively. HDAC removes acetyl groups from histones, and the resultant positive charges (from lysine and arginine) enhance interaction with negatively charged phosphate groups on the DNA molecule, which leads to reduced rate of transcription. Therefore, by inhibiting activity of HDAC, the SCFAs enhance transcription of certain beneficial anti-inflammatory genes and enable the cell to maintain homeostasis through regulation of the delicate balance between acetylation and deacetylation of cellular proteins. During infection, the phagocytotic activity of macrophages is important in disease resistance. Macrophages also produce reactive oxygen species (ROS) that are

toxic to microbes at high concentrations. It has been demonstrated that SCFAs can stimulate increased phagocytotic activity of macrophages, though other researchers have reported inhibition. The stimulation of ROS production by SCFAs has also not been completely proven as there are reports of inhibitory and stimulatory effects. The lack of consistency in the effects of SCFAs on macrophage activity could be due to differences in protocols used by various researchers. For example, experiments have reported differences in concentrations of SCFAs, source of macrophages (rat vs. human blood), methodology for ROS measurement, type of stimuli, and even pH of the measuring solution. Similarly, there is lack of a consensus regarding the actual beneficial effects of SCFAs that are produced from dietary factors such as fibers and prebiotics. However, animal experiments have shown beneficial effects of dietary intervention with RS, fiber, or cellobiose in inflammatory bowel disease. Such beneficial effects include reduced activities of myeloperoxidase (catalyzes production of tyrosine radicals), nitric oxide (NO) synthase (produces NO), reduction of weight loss, attenuation of inflammatory cytokine concentration, restoration of colonic glutathione (powerful antioxidant) levels, diminished tissue edema, epithelial cell proliferation, and enhanced growth of intestinal bacteria.

Bibliography

Arai, C., N. Arai, A. Mizote, K. Kohno, K. Iwaki, T. Hanaya, S. Arai, S. Ushio, and S. Fukuda. 2010. Trehalose prevents adipocyte hypertrophy and mitigates insulin resistance. *Nutrition Research* 30: 840–848.

Broekaert, W.F., C.M. Courtin, K. Verbeke, T. Van de Wiele, W. Verstraete, and J.A. Delcour. 2011. Prebiotic and other health-related effects of cereal-derived arabinoxylans, arabinoxylan-oligosaccharides, and xylooligosaccharides. *Critical Reviews in Food Science and Nutrition* 51: 178–194.

Brommage, R., C. Binacua, S. Antille, and A.-S. Carrie. 1993. Intestinal calcium absorption in rats is stimulated by dietary lactulose and other resistant sugars. *The Journal of Nutrition* 123: 2186–2194.

Franck, A. 2006. Oligofructose-enriched inulin stimulates calcium absorption and bone mineralization. *Nutrition Bulletin* 31: 341–345.

Gonzalez-Soto, R.A., L. Sanchez-Hernandez, J. Solorza-Feria, C. Nunez-Santiago, E. Flores-Huicochea, and L.A. Bello-Perez. 2006. Resistant starch production from non-conventional starch sources by extrusion. *Food Science and Technology International* 12: 5–11.

Kishida, T., H. Nogami, H. Ogawa, and K. Ebihara. 2002. The hypocholesterolemic effect of high amylose corn-starch in rats is mediated by an enlarged bile acid pool and increased fecal bile acid excretion, not by cecal fermented products. *The Journal of Nutrition* 132: 2519–2524.

Landing, K., G. Holm, L. Tengborn, and U. Smith. 1992. Guar gum improves insulin sensitivity, blood lipids, blood pressure, and fibrinolysis in healthy men. *American Journal of Clinical Nutrition* 56: 1061–1065.

Lattimer, J.M., and M.D. Haub. 2010. Effects of dietary fiber and its components on metabolic health. *Nutrients* 2: 1266–1289.

Lehmann, U., and F. Robin. 2007. Slowly digestible starch- its structure and health implications: a review. *Trends in Food Science and Technology* 18: 346–355.

Lin, Y., X.-F. Han, Z.-F. Fang, L.-Q. Che, D. Wu, X.-Q. Wu, and C.-M. Wu. 2012. The beneficial effect of fiber supplementation in high- or low-fat diets on fetal development and antioxidant defense capacity in the rat. *European Journal of Nutrition* 51: 19–27.

Liu, L., K.M. Winter, L. Stevenson, C. Morris, and D.N. Leach. 2012. Wheat bran lipophilic compounds with in vitro anticancer effects. *Food Chemistry* 130: 156–164.

Lopez, H.W., M.-A. Levrat-Verny, C. Coudray, C. Besson, V. Krespine, A. Messager, C. Demigne, and C. Remesy. 2001. Class 2 resistant starches lower plasma and liver lipids and improve mineral retention in rats. *The Journal of Nutrition* 131: 1283–1289.

Lopez-Molina, D., M.D. Navarro-Martinez, R.R. Melgarejo, A.N.P. Hiner, S. Chazarra, and J.N. Rodriguez-Lopez. 2005. Molecular properties and pre-biotic effect of inulin obtained from artichoke (*Cynara scolymus* L.). *Phytochemistry* 66: 1476–1484.

Olano, A., and N. Corzo. 2009. Lactulose as a food ingre-dient. *Journal of the Science of Food and Agriculture* 89: 1987–1990.

Rideout, T.C., Z. Yuan, M. Bakovic, Q. Liu, R.-K. Li, Y. Mine, and M.Z. Fan. 2007. Guar gum consumption increases hepatic nuclear SREBP2 and LDL receptor expression in pigs fed an atherogenic diet. *The Journal of Nutrition* 137: 568–572.

Roberfroid, M.B. 1998. Prebiotics and synbiotics: con-cepts and nutritional properties. *The British Journal of Nutrition* 80(Suppl. 2): S197–S202.

Scholz-Ahrens, K.E., P. Ade, B. Marten, P. Weber, W. Timm, Y. Asil, C.-C. Gluer, and J. Schrezenmeir. 2007. Prebiotics, probiotics, and synbiotics affect min-eral absorption, bone mineral content, and bone struc-ture. *The Journal of Nutrition* 137: 838S–846S.

Sharma, A., B.S. Yadav, and B.Y. Ritika. 2008. Resistant starch: physiological roles and food applications. *Food Reviews International* 24: 193–234.

Tzounis, X., J. Vulevic, G.C.C. Kuhnle, T. George, J. Leonczak, G.R. Gibson, C. Kwik-Uribe, and J.P.E. Spencer. 2008. Flavanol monomer-induced changes to the human faecal microflora. *The British Journal of Nutrition* 99: 782–792.

Vinolo, M.A.R., H.G. Rodrigues, R.T. Nachbar, and R. Curi. 2011. Regulation of inflammation by short chain fatty acids. *Nutrients* 3: 858–876.

Vuorinen-Markkola, H., M. Sinisalo, and V.A. Koivisto. 1992. Guar gum in insulin-dependent diabetes: effects on glycemic control and serum lipoproteins. *The American Journal of Clinical Nutrition* 56: 1056–1060.

Bioactive Lipids

2.1 Introduction

Several groups of lipids have been shown to provide health benefits either through modification of tissue fatty acid composition or induction of cell signaling pathways. While some health benefits are derived from consumption of short- to medium-chain fatty acids, evidence suggests that the polyunsaturated fatty acids (PUFAs) are the most important bioactive lipids. PUFAs are found mostly in plant seed oils and are important substrates for the biosynthesis of cellular hormones (eicosanoids) and other signaling compounds that modulate human health. The beneficial health effects of PUFAs seem to be dependent on their isomer configuration as the *cis*-isomer is the predominant bioactive form. Moreover, fatty acids in the *cis*-configuration have a rigid nonlinear structure, which enhances membrane fluidity when incorporated into cells. Increased membrane fluidity enhances cell to cell communication and helps maintain normal homeostasis or prevent the development of metabolic disorders. Therefore, this chapter will focus mostly on PUFAs but with brief discussions on a short-chain fatty acid (butyric acid) and medium-chain fatty acids. Detailed descriptions of the metabolic effects of short-chain fatty acids have been discussed in other parts of this book.

2.2 Butyric Acid

Butyrate is commonly found as part of the lipid component of dairy milk but is also one of the main by-products (others are acetate and propionate) of fiber fermentation in the colon and has been shown to induce various beneficial metabolic effects. Butyrate has been shown in vivo to be a stimulant of normal colonic cell proliferation but can also inhibit growth and proliferation of colon cancer cell lines. Other suggested health benefits of butyrate include:

- Substrate used for growth and regeneration of cells in large intestine.
- Anti-colon cancer properties probably through enhanced apoptosis of mutant colonic cells.
- Animal experiments also showed beneficial effect on the growth of cells in the small intestine.
- Increased thermogenesis to increase energy expenditure, which contributes to reduced body weight and other markers of metabolic syndrome.

R.E. Aluko, *Functional Foods and Nutraceuticals*, Food Science Text Series, DOI 10.1007/978-1-4614-3480-1_2, © Springer Science+Business Media, LLC 2012

2.3 Medium-Chain Fatty Acids

These are fatty acids that contain 8–10 carbon atoms, mainly caprylic (C8:0) and capric (C10:0) acids, which are metabolized differently when compared to long-chain fatty acids (14 or more carbon atoms). Medium-chain triglycerides (MCTs) contain medium-chain fatty acids (MCFAs) esterified to glycerol backbone and are usually completely hydrolyzed to yield the free fatty acids by lipases present in the gastrointestinal tract. When absorbed directly, MCTs enter the blood circulatory system through the portal vein and carried to the liver where they are oxidized to ketones. This is because in the mitochondria, transport of MCTs does not require carnitine palmitoyltransferase, a rate-limiting enzyme of β-oxidation. The mostly catabolic fate of MCFAs is evident by the fact that dietary MCTs reduce blood triglyceride levels during human intervention trials. Thus, dietary MCTs induce thermogenesis and do not contribute to weight gain since they are not deposited in the adipose tissue. This has been demonstrated in diet intervention trials involving hypertriglyceridemic human subjects where MCTs reduced body mass index, hip circumference, waist-hip ratio, total abdominal fat, visceral fat, body fat mass, and waist circumference. MCT diets also reduced blood levels of several types of LDL as well as LDL-cholesterol to greater extent than traditional oil that contained long-chain triglycerides. Therefore, MCTs may be used as a means of preventing and treatment of obesity, though the exact molecular mechanism of action has not been fully elucidated apart from the thermogenic effects. But MCTs were shown to activate hormone-sensitive lipase and down-regulate fatty acid synthase, which led to increased lipolysis and reduced fat accumulation, respectively, in white adipose tissue. And MCTs are able to upregulate expression of lipoprotein lipase, which is the major enzyme that is responsible for lipolysis. It should be noted that MCT oils are difficult to use as cooking oils because the presence of medium-chain fatty acids causes the oil to have lower smoke point than oils containing long-chain fatty acids. One solution to this problem has been the development of oils that contain triglycerides with combinations of MCFAs and long-chain fatty acids (LCFAs) esterified to the glycerol backbone; such oils are called medium- and long-chain triglycerides (MLCT), which have received regulatory approval in Japan for use in human foods. In type 2 diabetes, diet supplementation with 7% (w/w of diet) MLCT (contains ~13% MCFAs and ~87% LCFAs) led to significant decreases in mesenteric fat weight and postprandial insulin levels. Increase in mesenteric fat weight has been associated with insulin resistance; hence, the effect of MLCT has potential health benefits for blood glucose management. The MLCT diet-induced higher plasma levels of adiponectin was inversely correlated with mesenteric fat weight and plasma insulin level. The observed increased levels of plasma adiponectin may be because MCFAs can suppress adipocyte hypertrophy. Adiponectin is known to increase AMPK, an enzyme that increases muscle sensitivity to glucose uptake.

2.4 Long-Chain Fatty Acids

These are fatty acids with 14 or more linearly arranged carbon atoms and may be saturated (no double bonds) or unsaturated (one or more double bonds). These fatty acids are found mostly as components of the triglycerides of edible oils and fats.

2.4.1 Monounsaturated Fatty Acids

Feeding monkeys with a diet rich in oleic acid (60% of the fatty acids) led to up to 17% reduction in plasma total cholesterol concentrations when compared to the group that was fed a diet that was predominantly rich in saturated fatty acids.

Substitution of dietary monounsaturated for saturated fatty acids resulted in a 28% decrease in the Apolipoprotein B (ApoB) levels, which is due to lower production rates of LDL ApoB. There were also less amounts of circulating LDL

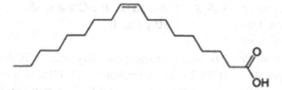

Fig. 2.1 Structural configuration of *cis*-oleic acid

particles in the plasma. Oleic acid (Fig. 2.1) has also been proposed to have a potential role in decreasing brain-related disorders such as dementia and Alzheimer's disease. Among unsaturated fatty acids tested, oleic acid had the highest in vitro inhibition of prolyl endopeptidase (PEP), an enzyme that is believed to have a role in amyloid formation in the brain. Affinities of fatty acids for PEP as measured by inhibition constants (inversely related to affinity) were ~27, 51, 89, 91, and 248 μM, respectively, for oleic, linoleic, docosahexaenoic, arachidonic, and eicosapentaenoic acids. PEP levels have been found to be upregulated in Alzheimer's disease patients, and rat experiments have shown improved cognitive functions when administered with PEP inhibitors. While in vivo studies are required to determine exact mechanisms of action, preliminary work suggests that consumption of oleic-acid-rich diets could have beneficial effects on brain functions by reducing activity of PEP.

Oleic acid has also been shown to have a potential role in the therapeutic management of colorectal cancer, one of the most common types of tumors, especially in western countries. The basic mechanism involves inhibition by oleic acid of the store-operated Ca^{2+} entry (SOCE) process that controls the Ca^{2+} influx pathway. Operation of the SOCE is believed to be involved in several cellular and physiological processes including cell proliferation. Therefore, attenuation of the SOCE process by oleic acid could reduce Ca^{2+} influx into cells and diminish or eliminate tumor cell proliferation. Oleic acid is able to block Ca^{2+} entry into cells, probably through free carboxylate-mediated metal chelation; this is because methylated oleic acid had no inhibitory effect on SOCE. Structural conformation of the fatty acid also seemed important contributory effect to inhibitory properties because

use of stearic acid (same carbon length but no double bond) had no effect on SOCE. Oleic acid blocks Ca^{2+} entry by binding to membrane molecules at the outer side of the membrane; therefore, it is possible that the structural conformation arising from the presence of a single double bond enhances interaction with the membrane surface. In contrast, conformation of the saturated stearic acid seems to be incompatible with the required binding protocol at the surface of the cell membrane.

2.4.2 Polyunsaturated Fatty Acids (PUFA)

Epidemiological studies have showed a low incidence of coronary heart diseases (CHD) in the Inuit population even though they consume a diet that is high in saturated fatty acid content. The low incidence of CHD was associated with the high levels of n-3 PUFAs that are also present in the mostly marine-product-laden diet of the Inuit. It is also known that a diet with high levels of linoleic and linolenic acids has better cholesterol-lowering effects when compared with a diet rich in saturated fatty acids. Unlike monounsaturated fats, a polyunsaturated-rich diet can decrease ApoB levels by a combination of reduced production and increased catabolism. ApoB is the main lipoprotein in low-density lipoproteins and is a marker of atherosclerosis and increased risk of cardiovascular damage because it is the main trafficker of cholesterol in the blood circulatory system. By decreasing blood circulating levels of ApoB, certain PUFAs can provide protection against certain cardiovascular diseases that arise from excess levels of vascular cholesterol. Increased dietary consumption of PUFAs is also associated with decreased blood levels of mediators of lipid-induced insulin resistance, increased insulin sensitivity, and enhanced leptin levels. The high plasma leptin levels reduces food intake due to appetite dampening effects. High levels of fish consumption (contains high PUFA levels) have been associated with improved immune response such as reduced risk of asthma-related symptoms and lower rate of allergic sensitization in addition to decreased levels of proinflammatory compounds such as C-reactive proteins, interleukin-6, and prostaglandins (PG). In fact, lack of fish consumption during childhood has been linked to increased risk of asthma development.

In general, long-chain PUFAs act as antihypertensive agents and reduce the risk for adverse cardiovascular events by enhancing production of vasodilatory PGs such as PG_1 and PG_2. For example, the metabolic products of linoleic acid (gamma-linolenic acid, GLA, and dihomo-GLA) are activators of PG_1 and PG_2 syntheses and have been found to prevent elevated blood pressure that is associated with consumption of saturated fatty acids. Eicosapentaenoic acid (EPA) and docosahexaenoic acid (DHA) are PUFAs that can reduce blood pressure and blood viscosity by enhancing formation of PGI_3 (a vasodilator and platelet anti-aggregator) and inhibiting thromboxane A2 (TXA_2, a potent vasoconstrictor and platelet aggregator). PUFAs also work as inhibitors of angiotensin-converting enzyme (ACE), a major enzyme responsible for increased formation of angiotensin II (a potent vasoconstrictor). The antihypertensive effects of PUFAs have also been shown to be associated with upregulation of endothelial nitric oxide production, and hypertensive patients have been shown to have low levels of PUFAs. The high level of PUFAs in human milk has been associated with reduced risk of developing hypertension in adulthood when compared to formula-fed infants that consume less amounts of PUFAs. By suppressing hypertension, PUFAs can also inhibit development of proteinuria (a marker of kidney damage) and prevent excessive proliferation of vascular smooth muscle cells through suppression of TGF-β synthesis. TGF-β is found in elevated concentrations in hypertensive patients, and interaction with angiotensin II leads to increased synthesis of extracellular matrix proteins within the kidney and aorta. Therefore, high levels of TGF-β promote renal scarring and pathological progression of end-stage renal disease in hypertensive and diabetic patients.

However, there are other various types of beneficial PUFAs, mostly the omega-3 (n-3) and omega-6 (n-6) fatty acids as well as the conjugated fatty acids.

2.4.3 Omega-3 and Omega-6 Fatty Acids

The simplest omega-6 fatty acid is linoleic acid (C18:2), while linolenic acid (C18:3) is the simplest omega-3 fatty acid. Both fatty acids have been reported to protect against cardiovascular and inflammatory diseases, though linolenic acid has greater health benefits. Typical examples of chronic diseases that have inflammation component and could benefit from increased dietary intake of omega-3 and omega-6 fatty acids include lupus, diabetes, psoriasis, obesity, Crohn's, rheumatoid arthritis, cystic fibrosis, Alzheimer's, and multiple sclerosis. While less effective than oleic acid, the omega fatty acids have been shown to reduce in vitro activity of PEP, an enzyme with potential role in the pathogenesis of brain diseases. It has been documented that blood levels of omega-3 fatty acids are inversely proportional to the risk of adverse cardiovascular events such as stroke and sudden death. In animal experiments, the ratio of omega-3 (n-3) to omega-6 (n-6) was an important determinant of ultimate health benefits. A higher ratio (more n-3 and less n-6) in the diet is more desirable as a means of improving human health such as reduced weight of intra-abdominal fat, adipocyte size, and normalization of heartbeat. This is because n-3 PUFAs are usually converted to anti-inflammatory eicosanoids while n-6 PUFAs are converted to proinflammatory eicosanoids. Thus, high levels of dietary n-3 PUFAs enhance the body's ability to reduce damaging inflammatory conditions that are known to be responsible for the initiation and growth of chronic diseases such as cancer, kidney malfunction, diabetes, and cardiovascular disorders. A human interventional trial involving >2,800 patients that survived a recent myocardial infarction (MI) showed that consumption of 1 g omega-3 fatty acid on a daily basis led to significant reduction in the cumulative rate of all-cause death and nonfatal MI. Specifically, combinations of docosahexaenoic acid (DHA) and eicosapentaenoic acid (EPA) with α-linolenic acid can lower the risk of fatal ischemic heart disease in older adults. Fish oil contains a high n-3:n-6 ratio and has been shown to decrease serum triglyceride and cholesterol

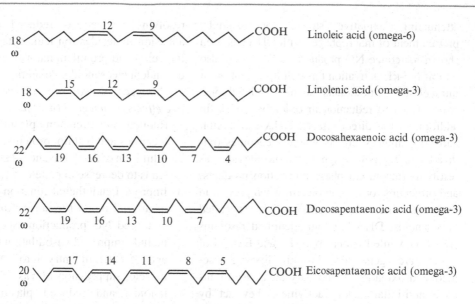

Fig. 2.2 Structural configurations of polyunsaturated fatty acids showing number of double bonds and position of the omega double bond (double bond nearest to the ω-carbon atom)

levels. However, during clinical trials, the total amount of fish oil consumed was a more important determinant of health benefits than the ratio of n-3 to n-6. Fish oil is also able to lower blood pressure in hypertensive patients. The health benefit of fish oil is due mostly to the n-3 PUFA level, especially EPA and DHA. α-Linolenic acid (ALA), which is present in vegetable sources such as rapeseed, walnuts, flaxseed, and green leafy products, may be desaturated and elongated within the human body to yield EPA, docosapentaenoic acid (DPA), and DHA. For example, daily consumption of 3 g ALA (given as flaxseed oil) resulted in 25% and 60% increases in plasma levels of DPA and EPA, respectively, but no effect on DHA level. Therefore, ALA may serve as a dietary precursor of EPA and DPA for vegetarians that do not consume fish oil products. However, ALA should not be used as the sole source of n-3 PUFAs since current evidence shows that it is only partially converted to EPA and DPA, but not DHA, especially in men. In women, partial conversion of ALA occurs probably because of less β-oxidation and role of estrogens in promoting fatty acid desaturation. The role of estrogens is evident from the fact that women using oral contraceptive pill that contains 17α-ethynyloestradiol had a threefold higher rate of DHA synthesis

when compared to women not on the pill. Testosterone is also known to decrease DHA synthesis, which may be responsible for the reduced conversion of ALA to DHA in men. However, this effect (conversion of ALA to DHA) is not seen in lactating women, and therefore, DHA content of breast milk is influenced mostly by the fat storage during and before pregnancy, which reinforces the need for adequate nutrition in women of childbearing age. The lack of ALA to DHA conversion during lactation may be due to the action of prolactin (lactating hormone), which is a known estrogen suppressor. Preformed DHA is a critical component of a healthy diet and is recommended as a means of maintaining physiologically beneficial levels. Figure 2.2 shows the chemical structures of the most common omega-3 fatty acids.

DHA also has anti-inflammatory effects through its inhibitory effects on activation of nuclear factor kappa-B (NF-κB). During high oxidative stress conditions within cells, the high levels of ROS activate NF-κB by releasing it from the bound inhibitory protein (IκB). Activated NF-κB then moves into the nucleus where it upregulates gene expression for several proinflammatory cytokines (IL-2, IL-6, and IL-8) and nitric oxide (NO). DHA acts as an inhibitor of NF-κB by attenuating oxidative stress; for example, pretreatment of macrophages with DHA has been shown to reduce NO production. DHA can also inhibit NF-κB activation through upregulation of intracellular antioxidants such as glutathione, which leads to reduction in oxidative stress. In addition to the direct effects, DHA can exhibit anti-inflammatory effects when it becomes oxidized to highly potent signaling molecules such as resolvins (resolution phase interaction products) and protectins (or neuroprotectin when present in the central nervous system), which are called docosanoids. DHA has four identified resolvins (D1–D4), while EPA has two (E1 and E2), all of which are generated through lipoxygenase-mediated oxidation of the fatty acids. Resolvins have anti-inflammatory activities; they act by downregulating activation of NF-κB and removal of neutrophils from inflammatory sites. Protectins are DHA-oxidative products formed by peripheral blood mononuclear cells and CD4 cells in response to oxidative stress conditions. Protectins are present in peripheral blood, lung tissues, neurons, and astrocytes. Neuroprotectins have beneficial health effects such as enhanced nerve regeneration and reduced leukocyte infiltration during inflammation. Generally, neuroprotectins help to maintain homeostasis during aging by reducing pro-apoptotic and proinflammatory signaling. As an antioxidative agent, synthesis of neuroprotectins is enhanced during oxidative stress to protect retinal and neuronal cells from apoptosis, probably through inhibition of IL-1β-induced induction of cyclooxygenases. These neuroprotective functions suggest that neuroprotectins (through increased dietary DHA) may be used for therapeutic management of neurodegenerative diseases, especially Alzheimer's as well as the normal cognitive decline associated with aging. Therefore, the anti-inflammatory effects of omega-3 fatty acids are mediated through various reactions such as reduced chemotactic responses of leukocytes and level of proinflammatory cytokines (via the NF-κB route), as well as increased EPA-dependent formation of weakly inflammatory or anti-inflammatory eicosanoids. Other effects include increased DHA- and EPA-dependent formation of anti-inflammatory and inflammation-resolving resolvins as well as reduced production of adhesion molecules (on leukocytes and endothelial cells) and proinflammatory arachidonic-acid-dependent eicosanoid mediators.

Proposed mechanisms for the cardioprotective effects of omega-3 (n-3) fatty acids:

1. Reduction in circulating plasma levels of triglycerides and bad lipoproteins.
2. Inhibition of thromboxane A2 synthesis, which leads to decrease in platelet aggregation.
3. Improved endothelial function (better control of arterial blood pressure) primarily through enhanced NO production. Secondary effects include improved vasodilation through blockage of calcium entry into vascular smooth muscle, suppression of vasoconstrictor prostanoids, and reduced plasma epinephrine level.
4. Upregulation and downregulation of genes involved in the synthesis of proteins responsible for lipid oxidation and lipid synthesis, respectively.
5. Prevention of arrhythmias and sudden death.
6. Decrease in plasma homocysteine levels, a known risk factor for cardiovascular diseases. Though mechanism is not fully understood, it is possible that omega-3 fatty acids modulate gene expression of enzymes involved in homocysteine metabolism. For example, omega-3 fatty acids are known to upregulate activity and mRNA expression of methionine adenosyl transferase (MAT), which increases cystathionine β-synthase activity that removes homocysteine from the methionine cycle.
7. DHA and EPA enhance NO (vasodilator) production by altering lipid composition such that endothelial NO synthase is displaced from its negative regulator (caveolin-1).

Some of the beneficial effects of omega-3 PUFAs are further discussed below.

(a) n-3 PUFAs and cardiac arrhythmias
   • Arrhythmia: irregular or abnormal heartbeat.
   • Most common fatal arrhythmia is known as ventricular fibrillation (VF).
   • Studies have shown that VF can be prevented by n-3 PUFAs in cultured animal heart cells.

(b) Proposed mechanisms of anti-arrhythmic effect of n-3 PUFAs

- Incorporation and modification of myocyte cell membranes by n-3 PUFAs resulting in modulation of membrane ion channels
- Prevention of high accumulation of intra-cellular calcium
- Production of antithrombotic eicosanoids, which reduces the potential for plaque formation
- Influence on cell signaling mediated through phosphoinositides
- Induction of different antioxidant enzymes, which reduces level of reactive oxygen species

(c) Effects of n-3 PUFAs on blood lipid profile and atherosclerosis

- In hypertriglyceridemic patients, dietary n-3 PUFAs reduced blood triglycerides (TG) by up to 28% after 2 weeks. Longer trials (up to 16 weeks) resulted in up to 47% reduction in blood TG.
- Lower lipoprotein cholesterol, though evidence is stronger in animal studies than in human clinical trials. However, there is increase in HDL cholesterol and lowering of total cholesterol content.
- Anti-inflammatory effect, which leads to reduced platelet aggregation and inhibition of atherogenesis. Modulation of platelet aggregation is mediated through the eicosanoid pathway and is not as a result of direct effect of fatty acids on platelets. Reduced expression of cell-membrane-bound adhesion molecules also contributes to reduced potential for atherogenesis.
- Though the effect on blood pressure is minimal, DHA may be a more effective hypotensive agent than EPA through augmentation of the NO-dependent endothelium-dependent vasodilation. However, increased fish consumption leads to high levels of EPA and DHA in the blood, which favors blood pressure reduction.

(d) Effects of omega-3 (n-3) and omega-6 (n-6) PUFAs on cancer

The most commonly diagnosed type of cancer among men is prostate cancer, which is believed to be associated with dietary factors. One of the dietary factors that have been associated with prostate cancer development is ratio of n-6/n-3 PUFAs. It is known from various in vitro and animal experiments that the two types of PUFA have different and opposite effects on cancer pathogenesis. This is because n-6 PUFAs (linoleic and arachidonic acids) promote tumor development while n-3 fatty acids (α-linolenic acid, EPA, and DHA) suppress tumor carcinogenesis. Evidence suggests that EPA and DHA have antiproliferative effects on cancer cells, but direct relationships to prostate cancer have been mixed with some researchers reporting negative while others reported positive associations. Some studies have shown that high levels of dietary linoleic acid (n-6 PUFA) are positively correlated with elevated risk of prostate cancer development. Most importantly, it is the balance of n-3 to n-6 PUFAs that is believed to be the main factor in tumor carcinogenesis. From epidemiological studies, it is known that high n-6/n-3 PUFA ratio is associated with prostate cancer risk in men, though the trend was dependent on race of the patient. In white men, high n-6/n-3 PUFA ratio was associated with risk of overall prostate cancer as well as risk of developing high-grade form of the tumor. However, in African-American men, high n-6/n-3 PUFA ratio was associated only with risk of developing high-grade form of prostate cancer tumor. The main mechanism for the beneficial effects of high dietary n-3 PUFAs on prostate cancer is believed to be through the competitive inhibition of conversion of n-6 PUFAs to proinflammatory eicosanoids. This is because n-3 and n-6 PUFAs compete for similar enzymes during eicosanoid synthesis; therefore, high levels of n-3 PUFAs will reduce catabolism of n-6 PUFAs but increase formation of anti-inflammatory eicosanoids. High oxidative state coupled with high levels of proinflammatory eicosanoids can cause damage to critical cellular components such as the DNA, which could lead to carcinogenesis.

By reducing the level of proinflammatory eicosanoids in prostate cells, the n-3 PUFAs have the potential to limit cellular damage and reduce the risk for carcinogenesis. Using adult mice colonocytes, it was shown that DHA in combination with butyrate was better than EPA/butyrate in inducing apoptosis.

Omega-3 fatty acid-containing fish oil has also been shown to induce apoptosis but confers resistance to oxidation-induced DNA damage in colonic cells. Using mice with insufficient activity of superoxide dismutase (SOD2), it was shown that fish oil can increase oxidative stress and lead to increase apoptosis of the colon cancer cells. SOD is one of the main antioxidant enzymes responsible for free radical scavenging; deficiency usually leads to increased mitochondria oxidative stress. In normal cells, high oxidative stress could cause health problems by damaging essential nutrients and cellular components. However, in abnormally growing cells like cancer cells, a high oxidative stress could be used to reduce growth and enhance apoptosis. This is because as the level of reactive oxygen species and lipoperoxides increases, eventually the mitochondria detoxification capacity is exceeded, and the resultant chronic oxidative stress triggers release of pro-apoptotic factors from the mitochondria into the cytosol. In this case, the high level of unsaturated omega-3 fatty acids coupled with the reduced SOD level can induce such pro-apoptotic condition of chronic oxidative stress.

(e) Omega-3 PUFAs, obesity, and kidney disease

Obesity continues to be an important risk factor for other chronic metabolic disorders such as insulin resistance (type 1 diabetes), hypertension, kidney function impairment, and cancer, all of which can lead to death. In a rat model (Han:SPRD-cy) of chronic kidney disease (CKD), supplementation of diet with omega-3-rich flaxseed oil (FO) resulted in increased whole-body bone mineral content (~1 g) and density (~0.5 g/cm2) as well as increased lean body mass (~12 g) in males when compared to omega-3-deficient control diet that contained corn oil. The FO was only effective in increasing whole-body bone mineral content (~0.5 g) and density (~0.05 g/cm2) in the female Han:SPRD-cy rats. Thus, it seems that the effect of the FO differed according to gender of the rats. The results are important because decreases in bone mineral density and lean body mass as well as increase in adipose tissue mass are associated with chronic renal failure and renal transplant. The Han:SPRD-cy is an autosomal-dominant inheritance that is characterized by epithelial proliferation, progressive dilatation of nephrons, interstitial inflammation, oxidative injury, and fibrosis. FO also reduced some of these renal disease markers such as numbers of macrophage cells and proliferating cell nuclear antigen (inflammation) as well as oxidized LDL (oxidative injury) content. Thus, FO may serve as a suitable therapeutic intervention tool to reduce the degree of renal inflammation, oxidative injury, and severity of lean body mass losses associated with CKD. It has also been demonstrated that dietary intervention through maternal nutrition may be effective in reducing pathological intensity or progression of genetically inherited CKD. For example, supplementation of adult female Han:SPRD-cy rat diets with FO led to a 15% decrease in renal cyst growth, 12% decrease in cell proliferation, and 15% decrease in oxidative injury in the offsprings that inherited the disease and maintained on FO-free diet. However, the renal health benefits of FO were increased when the offsprings from FO-fed mothers were also maintained on FO diet, postweaning. FO-fed Han:SPRD-cy offspring rats from FO-fed mothers had reduced proteinuria (13%), creatinine clearance rate (30%), renal fibrosis (34%), and glomerular hypertrophy (23%). FO also mitigated the detrimental effects of a high-fat diet on renal fibrosis in polycystic kidney disease mice. The mechanism involved in the renoprotective effects of FO may be due to modulation of eicosanoid synthesis in favor of less inflammatory agents. In FO-fed rats, the tissues, organs, and plasma contain higher levels of omega-3 fatty acids, especially α-linolenic acid (ALA), which is a precursor for synthesis of longer-chain omega-3 fatty acids, especially EPA. The high plasma concentrations of ALA inhibit conversion of linoleic acid to arachidonic acid (AA); therefore, concentration of EPA increases while that of AA decreases. EPA is converted slowly to the less vasoactive thromboxane A3, while AA is converted faster to thromboxane A2, a strong vasoconstrictor, and both fatty acids compete for the same metabolic (conversion) enzymes. Therefore, the high levels of EPA lead to competitive inhibition of AA conversion, which alters metabolic products in favor of EPA-derived eicosanoids and reduction in renal injury. Thus, FO may be used in maternal diet to alter eicosanoid production during pregnancy, which can then attenuate disease symptoms associated with inherited CKD in the offsprings.

Various experiments using nondiabetic animals have confirmed the potential use of omega-3 fatty acids as therapeutic agents to reduce plasma triglycerides and prevent (or treat) excessive accumulation of body fat and even weight gain. In diet-induced obesity, omega-3-supplemented rat diets led to a decrease in epididymal fat in addition to attenuation of the increase in retroperitoneal fat mass, which were attributed to reductions in number of mature adipocytes and adipocyte hypertrophy but not adipocyte number. It is important to note that anti-obesity of omega-3 oil-supplemented diets may be due to the content of DHA because low ratio of EPA/DHA has been shown to promote reduced accumulation of subcutaneous fat. However, canola oil has also been shown to reduce accumulation of intra-abdominal fat mass, which was associated with less adipocyte surface area. Human studies have also shown that abdominal obesity and visceral abdominal fat area are inversely related to omega-3 fatty acid content (especially DHA) of perivisceral and omental adipose tissues.

Similarly, the size of adipocyte present in the subcutaneous adipose tissue was inversely related to the content of omega-3 fatty acids. In non-obese, healthy adults with BMI of 20–40 kg/m², plasma omega-3 fatty acid levels were found to be inversely proportional to waist and hip circumferences as well as BMI, which indicates a protective role against obesity for this group of fatty acids. However, supplementation of diabetic mice diet actually exacerbated weight gain, suggesting that impaired glucose control may nullify the weight-reducing effects of omega-3 fatty acids. Apart from weight gain prevention, there has been a limited study on the role of omega-3 fatty acids in established obesity. Dietary omega-3 fatty acid was shown to reduce body fat mass in mice that were made obese through intake of high-fat diet. The loss in weight was attributed to reduced metabolic efficiency because of the decreased food efficiency (lowest weight gain per unit energy intake) in obese mice that whose diet was switched to omega-3-supplemented feed. In humans, the mechanism involved in omega-3 fatty acid-induced weight loss has been shown to involve decreased appetite, modulation of lipogenic gene expression, and tissue metabolism. Human subjects on omega-3-supplemented diets have been shown to consume less amount of food than equivalents on control diets. It is believed that omega-3 fatty acids can increase postprandial satiety in overweight and obese individuals, which reduces food intake and calories and which enhances body weight loss and improved body composition. Increased β-oxidation of fatty acids through upregulation of mitochondrial carnitine palmitoyl transferase 1 (CPT-1) has been shown to be associated with dietary omega-3 fatty acids. CPT-1 exchanges coenzyme A for carnitine, which then facilitates movement of fatty acids into the mitochondria for β-oxidation. Mitochondrial expression of CPT-1 is regulated by peroxisome proliferator-activated receptors (PPARs) and by AMP-activated protein kinase (AMPK). AMPK is activated by EPA in adipose tissue and skeletal muscle, which then leads to upregulation of CPT-1 expression and increased fatty acid oxidation in the mitochondria. This metabolic regulation of fatty acid oxidation has been demonstrated in rat feeding experiments where supplementation of the diet with EPA-rich fish (menhaden) oil led to significant increase in skeletal muscle mitochondrial CPT-1 activity and reduced sensitivity to inhibitors when compared to low EPA types of oils. The lower sensitivity of CPT-1 ensures increased lipid oxidation even in the presence of other dietary factors that may act as inhibitors.

Another mechanism proposed for the adipose tissue effects (decreased fat deposition) of omega-3 fatty acids is through increased expression of uncoupling protein 3 (UCP-3) mRNA in the skeletal muscle coupled with increased expression of peroxisomal acyl-CoA oxidase (PACO) in liver, heart, and skeletal muscle. However, the PACO pathway is less efficient for energy production because it produces 30–40% more heat and 30% less ATP when compared to mitochondrial β-oxidation. The UCP-3 reduces mitochondrial oxidative phosphorylation efficiency because it induces leakage of protons from the mitochondria, which leads to less ATP formation but more heat production. Thus, the combined effects of PACO and UCP-3 lead to overall reduction in metabolic efficiency and contributes to decreased stored fat energy (due to decreased availability of fatty acids) but increase energy losses in the form of heat. The ability of omega-3 fatty acids to reduce adipose tissue weight can also be linked to upregulation of intestinal lipid oxidation. This is because dietary omega-3 fatty acids have been shown to increase intestinal expression of mRNAs for lipid-oxidizing agents such as CPT-1a, cytochrome P450 4A10, and malic enzymes. In addition to the indirect effects on skeletal muscle, intestinal tissue, heart, and liver, there is evidence for the direct effects of omega-3 fatty acids on adipose tissue in the form of increased fat oxidative and decreased fat synthesis capacities in visceral fat depots. The increased fat oxidation in visceral fats resulted from upregulated levels of peroxisome proliferator-activated receptor gamma coactivator 1α (PGC-1α) and nuclear respiratory factor-1 (NRF-1) that regulate mitochondria biogenesis, as well as CPT-1 that regulates fatty transfer into the mitochondria for β-oxidation.

There was also reduced mRNA expression level of stearoyl-CoA desaturase (a lipogenic enzyme), which led to reduced fat synthesis that was associated with the omega-3-supplemented diet. It has been hypothesized that the increased flow of fatty acids into the skeletal muscle for mitochondrial oxidation could have been due to the fact that omega-3 fatty acids enhance vasodilation and blood flow. The increased vascular blood flow enhances nutrient delivery to the skeletal muscles, where the nutrients are utilized for energy production, but reduces availability of nutrients for fat synthesis and storage in the adipose tissue.

Animal studies have shown the ability of omega-3 fatty acids to alter the metabolic pathways in skeletal muscle and promote protein synthesis to maintain lean muscle mass. Increased protein synthesis required higher metabolic rate and fatty acid utilization (to produce ATP), which could indirectly contribute to reductions in adipose tissue mass. For example, EPA is a known indirect suppressor of the ubiquitin-proteasome pathway that is critical for muscle proteolysis to occur. This is because EPA attenuates activation of the transcription factor, NF-κβ, a positive modulator of the ubiquitin-proteasome pathway. Increased dietary omega-3 levels promote muscle protein synthesis by activating key protein synthesis regulatory kinases such as mammalian (mechanistic) target of rapamycin (mTOR) and S6K. While the specific effects of DHA have been reported, the anti-obesity activity of individual omega-3 fatty acids is not fully elucidated. Therefore, the observed beneficial effects of omega-3 oils on weight reduction could be due to summation of individual effects or synergistic interactions. Overall, the net effect of the inhibition of protein degradation and promotion of protein synthesis is decreased substrate for building adipose tissue mass and hence reduced weight gain.

Potential adverse effects of omega-3 fatty acids

1. At very high dietary doses (>20 g/day) of omega-3 fatty acids, there is the potential for increased bleeding times that is not seen with moderate (2–5 g/day) consumption.

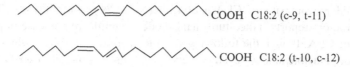

COOH C18:2 (c-9, t-11)

COOH C18:2 (t-10, c-12)

Fig. 2.3 Structures of the two main forms of conjugated linoleic acid (CLA) present in food product

2. High dietary doses may also lead to increase (<5%) in plasma LDL concentration. It is recommended that administration of an HMG-CoA reductase inhibitor (e.g., statins) could be used to offset the increase in LDL concentration. Inhibition of HMG-CoA reductase leads to upregulated expression of liver LDL receptors, which enhances uptake of LDL from the plasma and subsequent catabolism.

3. In some people, consumption of oils or foods that are rich in omega-3 fatty acids may induce gastrointestinal discomfort (bloating, belching, stomach upset) and nausea.

2.4.3.1 Conjugated Linoleic Acid (CLA)

CLA is a mixture of positional and geometrical isomers of linoleic acid formed by rumen microorganisms. CLA occurs naturally and can be found at low levels in ruminant fats such as beef tallow and milk fat. CLA can also be synthesized from linoleic acid or vegetable oils that have high levels of linoleic acid such as corn, canola, soybean, safflower, and sunflower. The principal bioactive dietary CLA is *cis(c)*-9, *trans(t)*-11 isomer, which is present at 73–94% content of the total CLA in milk, dairy products, meat, and processed meat products of ruminant origin. The following 17 natural CLA isomers have been detected in food products, though *c*9, *t*11 and *t*10, *c*12 isomers (Fig. 2.3) are the predominant forms found in food products:

*c*9, *t*11; *t*10, *c*12; *c*9, *c*11; *c*7, *t*9; *c*11, *t*13; *c*8, *t*10; *c*10, *t*12; *c*11, *c*13; *c*12, *t*14; *t*9, *t*11; *t*8, *t*10; *t*7, *t*9; *t*7, *c*9; *t*12, *t*14; *t*11, *t*13; *t*6, *t*8; *t*11, *c*13

The main structural features and potential physiological functions of CLAs are as follows:

- In linoleic acid, the two double bonds are separated by two single bonds:

  - CH = CH - CH2 - CH = CH -

- In conjugated linoleic acid, the two double bonds are separated by one single bond:

  - CH = CH - CH = CH -

- CLA is an essential fatty acid found mostly in animal products such as milk and meat.

- Meat of ruminants – cows, sheep, and other animals – that chew the cud contains more CLA than nonruminant meats such as turkey, chicken, and pork.

- This is because bacteria in the stomach of ruminants convert linoleic acid to CLA, which is absorbed into the animal tissue.

- Nonruminants do not have this type of bacteria, so they cannot produce CLA.

- Has been found to suppress atherosclerosis (plaque formation that leads to heart attack) in blood vessels.

- Potential anticarcinogenic effect has been shown in rats.

- A year 2000 survey determined that women with the most CLA in their diets had a 60% reduction in the risk of breast cancer.

- Has been shown to reduce fat mass and preserve muscle tissue in rats. Human benefit for weight loss has not been completely demonstrated.

- Lowers insulin resistance which may help prevent adult-onset of diabetes.

- Enhances immunity and resistance to infections.

- Milk fat contains 5–7 mg/g of CLA.

- CLA has been shown to inhibit growth of tumors in experimental rats.

- The 9-cis,11-trans isomer is believed to be the most biologically active.

- Tumor growth inhibition may be due to ability of CLA to inhibit protein and nucleotide biosynthesis.

Proposed mechanisms of CLA action: there are contradictory reports in literature on the mode of action of CLA, though the following seem to be the most plausible among several others:

- Indirect antioxidant property through direct scavenging of free radicals and inhibition of lipid peroxidation. CLA also upregulates vitamin E level, a potent antioxidant. Thus, CLA protects membranes and tissues from destructive oxidative stress by maintaining structural integrity of essential fatty acids, which also enhances membrane fluidity.
- Inhibition of carcinogen-DNA adduct formation.
- Induction of apoptosis.
- Modulation of tissue fatty acid composition and eicosanoid metabolism; CLA inhibits activities of lipoxygenase (leads to reduced formation of leukotrienes) and cyclooxygenase (leads to reduced formation of prostaglandins). This leads to reductions in signal transduction and cellular activities.
- Inhibition of hepatic 3-hydroxyl 3-methyl glutaryl CoA (HMG-CoA) reductase activity, which attenuates cholesterol synthesis with beneficial effects on plasma cholesterol content.
- Increases activity of enzymes involved in fatty acid oxidation (e.g., carnitine palmitoyl transferase) while decreasing fatty acid synthesis through inhibition of fatty acid synthase activity.
- Affects expression and action of cytokines and growth factors.
- Regulation of certain nuclear receptors involved in the control of body weight and adiposity either by reducing level of expression or translational ability of the genes that code for these receptors.

However, the potential modulation of chronic kidney disease by CLA isomers has received considerable attention, and promising effects have been demonstrated. In adult male Han:SPRD-*cy* rats with advanced kidney disease, diet supplementation with 1% CLA mixture (52% c9, t11: 3% t10, c12: 40% other geometrical isomers) led to significant decreases in oxidative damage (30%), proliferating cells (28%), inflammation (42%), and fibrosis (28%). There was also a significant decrease in production of parathyroid hormone (PTH), which is normally elevated in this rat model of CKD. PTH is known to induce bone loss during kidney disease. However, the CLA diet did not produce any significant effect on renal function, which suggests that reductions in inflammation and oxidative damage markers alone may not be sufficient to prevent deterioration of renal functions. In obesity-associated kidney disease, dietary CLA reduced pathological symptoms such as kidney weight (7%), glomeruli size (20%), and COX-2 protein levels (39%). In the obese rats, dietary CLA was also effective in preserving pancreatic islets, improving peripheral utilization of glucose, and reducing level of inflammatory agents. CLA reduced adipocyte size, hepatic steatosis, urinary albumin, and plasma lipids, but liver function was improved.

2.4.3.2 Conjugated Eicosapentaenoic Acid (CEPA)

CEPA can be prepared by alkaline treatment of eicosapentaenoic acid (EPA) and has been shown to induce strong and selective in vitro apoptosis of tumor cells through a lipid peroxidation mechanism. In animal experiments, CEPA had stronger antitumor effects, especially by preventing development of new blood vessels (angiogenesis) when compared to CLA and EPA. Through this mechanism, CEPA acts by cutting off the flow of nutrients into the developing tumor cells; the malnourished cells will eventually die off and scavenged by macrophages. In order to prevent angiogenesis, CEPA inhibits secretion and mRNA expression of matrix metalloproteinases (MMP), in particular MMP2 and MMP9 as shown in tissue culture experiments. This is because MMP are a group of enzymes that cause degradation of sub-endothelial basement membrane and surrounding extracellular matrix, which is then followed by migration and proliferation of the endothelial cells to form new vessels. Since MMP are key factors in angiogenesis, their inhibition by CEPA limits cell migration and provides a therapeutic approach to preventing growth and metastasis of tumors. It has also been suggested that CEPA may act as an antitumor agent by promoting increased lipid peroxidation in tumor cells. These tumor cells are known to have less antioxidant defense than normal cells; therefore, CEPA can cause accumulation of high levels of lipid peroxides, which eventually becomes toxic and lead to apoptotic cell death in the tumor. This is evident by the fact that CEPA had no effect on the proliferation of normal cells. Another potential mechanism for the anticancer activity of CEPA is through direct inhibition of topoisomerases, enzymes that are involved in DNA replication. Using in vitro methods, it has been shown that CEPA binds to topoisomerases I and II, which prevents interaction of the enzymes with DNA strands and hence reduced ability for cell division and proliferation. CEPA also caused arrest of cell replication at the G1/S-phase and prevented incorporation of thymidine into the cells, thus blocking the primary step of DNA replication by inhibiting activity of DNA polymerases. The arrest of G1/S-phase cell replication is due to CEPA-dependent enhanced levels of cyclins A and E; excessive cyclin levels may be due to inhibition of polymerase activity by CEPA. These anti-replication effects of CEPA led to increased apoptosis of the tumor cells.

CEPA has also been investigated for potential anti-obesity effects with promising data on total lipid reduction and attenuated adipose tissue growth. In rats, dietary CEPA was associated with reduced body weight and epididymal adipose tissue mass, a visceral white adipose tissue. Plasma levels of free fatty acid (FFA), triglycerides, total cholesterol, and TNF-α (a known inducer of insulin resistance) were significantly lowered by dietary treatment of rats with CEPA. Adipocytes grow in size by increasing intracellular accumulation of lipids in addition to secreting large amounts of TNF-α and FFA; therefore, CEPA reduced adipose tissue weight by attenuating cellular activities of adipocytes. Analysis of hepatic enzymes showed that rats that consumed CEPA diets had higher levels of lipid-catabolizing enzymes but reduced levels of lipid-synthesizing enzymes. For example, activity of fatty acid synthase and malic acid was significantly reduced, while activity of acyl-CoA oxidase (rate-limiting enzyme in hepatic fatty acid β-oxidation) was

significantly increased. However, further tests with humans are required to confirm efficacy, required dosage, and safety levels of CEPA with respect to obesity prevention.

Bibliography

Bradbury, J. 2011. Docosahexaenoic acid (DHA): an ancient nutrient for the modern human brain. *Nutrients* 3: 529–554.

Brousseau, M., A.F. Stucchi, D.B. Vespa, E.J. Schaeffer, and R.J. Nicolosi. 1993. A diet rich in monounsaturated fats decreases low density lipoprotein concentrations in cynomolgus monkeys by a different mechanism than does a diet enriched in polyunsaturated fats. *The Journal of Nutrition* 123: 2049–2058.

Buckley, J.D., and P.R.C. Howe. 2010. Long-chain omega-3 polyunsaturated fatty acids may be beneficial for reducing obesity- a review. *Nutrients* 2: 1212–1230.

Calder, P. 2010. Omega-3 fatty acids and inflammatory processes. *Nutrients* 2: 355–374.

Carrillo, C., M. del Mar Cavia, and S.R. Alonso-Torre. 2012. Oleic acid inhibits store-operated calcium entry in human colorectal adenocarcinoma cells. *European Journal of Nutrition*. doi:10.1007/s00394-011-0246-8.

Das, U.N. 2004. Long-chain polyunsaturated fatty acids interact with nitric oxide, superoxide anion, and transforming growth factor-β to prevent human essential hypertension. *European Journal of Clinical Nutrition* 58: 195–203.

Dhiman, T.R., S.-H. Nam, and A.L. Ure. 2005. Factors affecting conjugated linoleic acid content in milk and meat. *Critical Reviews in Food Science and Nutrition* 45: 463–482.

Fan, Y.-Y., Y. Zhan, H.M. Aukema, L.A. Davidson, L. Zhou, E. Callaway, Y. Tian, B.R. Weeks, J.R. Lupton, S. Toyokuni, and R.S. Chapkin. 2009. Proapoptotic effects of dietary (n-3) fatty acids are enhanced in colonocytes of manganese-dependent superoxide dismutase knockout mice. *The Journal of Nutrition* 139: 1328–1332.

Huang, T., J. Zheng, Y. Chen, B. Yang, M.L. Wahlqvist, and L. Duo. 2011. High consumption of Ω-3 polyunsaturated fatty acids decrease plasma homocysteine: a meta-analysis of randomized, placebo-controlled trials. *Nutrition* 27: 863–867.

Liu, Y., C. Xue, Y. Zhang, Q. Xu, X. Yu, X. Zhang, J. Wang, R. Zhang, X. Gong, and C. Guo. 2011. Triglyceride with medium-chain fatty acids increases the activity and expression of hormone-sensitive lipase in white adipose tissue of C57BL/6J mice. *Bioscience, Biotechnology, and Biochemistry* 75: 1939–1944.

Mata Lopez, P., and R.M. Ortega. 2003. Omega-3 fatty acids in the prevention and control of cardiovascular disease. *European Journal of Clinical Nutrition* 57(Suppl. 1): S22–S25.

Park, Y.-S., H.-J. Jang, K.-H. Lee, T.-R. Hahn, and Y.-S. Paik. 2006. Prolyl endopeptidase inhibitory activity of unsaturated fatty acids. *Journal of Agricultural and Food Chemistry* 54: 1238–1242.

Sankaran, D., N. Bankovic-Calic, C.Y.-C. Peng, M.R. Ogborn, and H.M. Aukema. 2006. Dietary flax oil during pregnancy and lactation retards disease progression in rat offspring with inherited kidney disease. *Pediatric Research* 60: 729–733.

Sankaran, D., N. Bankovic-Calic, L. Cahill, C.Y.-C. Peng, M.R. Ogborn, and H.M. Aukema. 2007. Late dietary intervention limits benefits of soy protein or flax oil in experimental polycystic kidney disease. *Nephron. Experimental Nephrology* 106: e122–e128.

Siddiqui, R.A., K.A. Harvey, and G.P. Zaloga. 2008. Modulation of enzymatic activities by n-3 polyunsaturated fatty acids to support cardiovascular health. *The Journal of Nutritional Biochemistry* 19: 417–437.

Terada, S., S. Yamamoto, S. Sekine, and T. Aoyama. 2012. Dietary intake of medium- and long-chain triacylglycerols ameliorates insulin resistance in rats fed a high-fat diet. *Nutrition* 28: 92–97.

Tsuzuki, T., Y. Kawakami, Y. Suzuki, R. Abe, K. Nakagawa, and T. Miyazawa. 2005. Intake of conjugated eicosapentaenoic acid suppresses lipid accumulation in liver and epididymal adipose tissue in rats. *Lipids* 40: 1117–1123.

Tsuzuki, T., A. Shibata, Y. Kawakami, K. Nakagawa, and T. Miyazawa. 2007. Conjugated eicosapentaenoic acid inhibits vascular endothelial growth factor-induced angiogenesis by suppressing the migration of human umbilical vein endothelial cells. *The Journal of Nutrition* 137: 641–646.

Weiler, H.A., H. Kovacs, E. Nitschmann, N. Bankovic-Calic, H. Aukema, and M. Ogborn. 2007. Feeding flaxseed oil but not secoisolariciresinol diglucoside results in higher bone mass in healthy rats and rats with kidney disease. *Prostaglandins, Leukotrienes, and Essential Fatty Acids* 76: 269–275.

Williams, C.M., and G. Burdge. 2006. Long-chain n-3 PUFA: plant v. marine sources. *The Proceedings of the Nutrition Society* 65: 42–50.

Williams, C.D., B.M. Whitley, C. Hoyo, D.J. Grant, J.D. Iraggi, K.A. Newman, L. Gerber, L.A. Taylor, M.G. McKeever, and S.J. Freedland. 2011. A high ratio of dietary n-6/n-3 polyunsaturated fatty acids is associated with increased risk of prostate cancer. *Nutrition Research* 31: 1–8.

Xue, C., Y. Liu, J. Wang, R. Zhang, Y. Zhang, J. Zhang, Y. Zhang, Z. Zheng, X. Yu, H. Jing, N. Nosaka, C. Arai, M. Kasai, T. Aoyama, and J. Wu. 2009. Consumption of medium- and long-chain triacylglycerols decreases body fat and blood triglyceride in Chinese hypertriglyceridemic subjects. *European Journal of Clinical Nutrition* 63: 879–886.

Bioactive Peptides

3.1   Introduction

These are short-chain protein molecules (usually <20 amino acid residues) that when ingested can provide physiological benefits such as reduction in the risk of cardiovascular diseases, negative modulation of tumor growth, and reduction in blood sugar level. Peptides are similar to proteins in being composed of amino acids but differ in having smaller number of such residues than proteins. In general, the average molecular weight of peptides is <10 kDa while proteins have sizes bigger than 10 kDa. Most bioactive peptides occur as amino acid sequences found within polypeptide chains of food proteins. When present as part of the native polypeptide sequence, the peptide sequences do not exhibit physiological activities and are known to be inactive. However, upon release by enzymatic or chemical hydrolysis, the free peptides become physiologically active (bioactive) and may be used as ingredients to formulate therapeutic foods. As shown in Fig. 3.1, various bioactive properties have been attributed to food protein-derived peptides, including blood pressure reduction, scavenging/neutralization of oxidative compounds, blood lipid/cholesterol reduction, enhanced calcium/mineral absorption, inhibition of microbial growth, and prevention of tumor formation. Several protein hydrolysates and peptides have also been identified to possess more than one of these attributes and are called multifunctional peptides. For example, some peptides have antihypertensive, antioxidant, and anticancer attributes.

Most bioactive peptides that are suitable for the formulation of therapeutic foods are usually less than 1 kDa (dipeptides to heptapeptides) but could be as big as 3 kDa in size. Low molecular weight peptides are desirable because they can be resistant to enzymatic hydrolysis during passage through the gastrointestinal tract (GIT). The small size of the peptides also enhances their absorption in intact form into the blood circulatory system from where they are transported to various target organs. In contrast, high molecular weight peptides will be possibly degraded by digestive enzymes during passage through the gastrointestinal tract; in most cases, structural degradation will render the peptide less active or even inactive. However, in a few cases, the peptides may be converted into more absorbable or active form during passage through the GIT. Though chemical hydrolysis is possible, enzymatic digestion with proteases is preferred for producing bioactive peptides that will be used to formulate functional foods and nutraceuticals. This is because chemical digestion with acid or alkali reagents can lead to production of unwanted products that can have negative effects on human health. In contrast, the enzymatic process produces well-defined peptide profiles with high degree of productivity and no undesirable by-products. Moreover, the chemical process can

R.E. Aluko, *Functional Foods and Nutraceuticals*, Food Science Text Series,
DOI 10.1007/978-1-4614-3480-1_3, © Springer Science+Business Media, LLC 2012

Fig. 3.1 Bioactive
properties of food
protein-derived peptides
relevant to the promotion
of human health and
disease prevention
(Adapted from Udenigwe
and Aluko (2012), with
permission of John Wiley
& Sons, Inc)

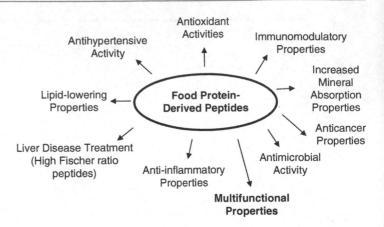

destroy the L-form of amino acids, which is the natural form that is required for nutritional and physiological activities of the peptides. Chemical hydrolysis can also produce potentially toxic or unwanted by-products such as lysinoalanine and D-form of amino acids. The use of chemical reagents can lead to a product with high residual levels of acid or alkali, which could preclude utilization of the protein digest in food products while subsequent neutralization could lead to unacceptably high levels of salt, usually in the form of NaCl. This is because the most commonly used alkali is NaOH while HCl is the most commonly used acid.

3.2   How to Produce Bioactive Peptides

There are three main strategies or approaches that can be used to produce bioactive peptides:

1. *Isolation from* in vitro *enzymatic digests of proteins.* In this approach, enzymes (usually of microbial or plant origin, e.g., alcalase, flavourzyme, papain, ficin, thermolysin, pronase, and neutrase) are used singly or as a combination of two or more enzymes to cleave large polypeptides (food proteins) into small peptides. During hydrolysis, peptide bonds are cleaved and there is increase in charge density as well decrease in molecular size, both of which contribute to increased solubility of the peptide products. Therefore, the peptides are more soluble than the native protein and centrifugation will clarify the reaction mixture; the precipitate contains undigested materials while the supernatant contains the soluble peptides. The supernatant is then separated and freeze-dried or used for further processing such as ultrafiltration. This method is the most commonly used in the production of bioactive peptides mostly because it is convenient, cheap, and can be performed easily in virtually any laboratory since it does not require expensive instrumentation. Alternatively, the native protein can be subjected to microbial fermentation during which the microbial enzymes act to hydrolyze the proteins into soluble peptides that can be recovered in the supernatant after centrifugation. Milk is the most commonly used substrate for microbial fermentation though fermentation of soybean has also led to the production of bioactive peptides.

2. *Isolation from actual or simulated* in vivo *GIT digests of precursor proteins.* Actual in vivo GIT digestion of proteins is possible, but it is rarely used for producing bioactive peptides because it requires removal of intestinal contents from live animals that have been fed a protein diet. Moreover, availability of enzymes that simulate GIT digestion as well availability of low-cost microbial enzymes makes in vivo production in a live animal an unattractive method. However, the GIT digestion may be simulated in vitro, first by treating the protein with pepsin for 1–2 h to simulate digestion in the stomach. The pepsin digest is then adjusted to a pH value around neutrality (usually pH 6.8–7.5), and

simulated intestinal digestion carried out using pancreatin (a pancreas extract that contains digestive enzymes commonly found in the intestine) or a combination of trypsin and chymotrypsin (the two most abundant digestive proteases in the intestine). This method is useful to estimate the potential bioactive peptides that can be liberated from a food protein when it is consumed as part of the diet, though it does not simulate potential action of intestinal brush border cellular proteases.

3. *Chemical synthesis of peptides having identical or similar structures to those known to be bioactive.* In most cases, the use of fractionation and purification methods does not produce adequate or economically feasible quantities of peptides. Therefore, the peptide sequences must be chemically synthesized. This approach can only be used if information on the amino acid sequence of known bioactive peptides is available, but it is a more cost-effective approach than actual peptide purification of peptides from protein digests. Chemical synthesis of peptides requires specialized instruments, which may not be available in most laboratories. Therefore, peptide synthesis is usually carried out by chemical companies on a fee for service basis. However, one of the main problems with chemical synthesis is the presence of chemical impurities and differences in stereochemistry that could affect activity when compared to the isolated natural peptide.

3.3 In Vitro Enzymatic Hydrolysis of Proteins

This method involves cell-free systems as well as microbial-based fermentation systems, utilizes food proteins as raw materials, and is the most common method for producing bioactive food protein digests.

3.3.1 Cell-Free System

Production of bioactive peptides generally begins with the selection and addition of appropriate amounts of a protease to the protein raw material, which could be of animal or plant sources. Typical examples of protein raw materials include casein, whey proteins, defatted seed meals, defatted animal products (skin, bones, feathers), seed protein concentrates, seed protein isolates, leaf protein concentrates, and mushroom proteins. As indicated above, enzymatic hydrolysis can be carried out with a single protease or a combination of more than one enzyme usually in a consecutive manner after the pH and temperature of the reaction mixture have been adjusted to levels that match the optimum working conditions of each enzyme. If the optimum pH and temperature for two or more enzymes are similar, they could be added simultaneously to the protein raw material as digestive agents to produce bioactive peptides. During digestion, the pH and temperature must be kept constant to maintain optimum enzyme activity. This is because protein digestion breaks peptide bonds and release H^+ that causes decreases in pH value of the reaction mixture. To maintain desired pH value, the reaction mixture is constantly monitored with a pH meter, and manual addition of alkali can be used to neutralize protons that are released as a result of protein hydrolysis. Automated pH stabilization can also be carried out using a software-driven instrument called the pH-stat, which enables protein digestion without the need for constant monitoring by an attendant. Depending on susceptibility of proteins to the digestive enzyme being used, most of the long-chain polypeptide chains will be converted into an assortment of various short-chain peptides that are usually more soluble than the native proteins. Therefore, at the end of digestion, the reaction mixture will consist of soluble short-chain peptides and insoluble native or long-chain polypeptides that were not digested. At the end of the set time period, protease activity can be stopped using three main methods if further protein digestion is not required:

(a) Adjusting the reaction mixture to a pH value where the enzyme is inactive. For example, if the enzyme is active in the range of pH 6–9, the reaction mixture will be adjusted to pH values less than 6. For most proteins, it is desirable to adjust the digest to the pI, i.e., isoelectric point (if it is outside the optimum protease activity range). This is because at the pI, the undigested proteins are insoluble and can be more readily separated from the soluble peptides. It is not recommended to adjust the digest to high pH values (higher than 8.0) because a combination of high temperature and pH could have adverse effects on the integrity of the amino acids present in target peptides. Moreover, at alkaline pH values, the undesirable undigested proteins might become solubilized and become difficult to separate from the desirable peptides.

(b) The reaction mixture can be heated to temperature levels that cause enzyme denaturation and coagulation of undigested proteins. Most enzyme proteins are denatured at temperatures above 70°C; therefore, maintaining the reaction mixture at or above this level for 10–20 min inactivates the enzyme and stops proteolysis. For thermophilic proteases, higher temperatures may be required for inactivation or method "c" below can be used.

(c) A combination of appropriate pH adjustment and heat application.

After the enzyme reaction has been stopped, the digested mixture is cooled and centrifuged at high g-force (usually $>10,000 \times g$) to obtain a clear supernatant solution and a precipitate. Soluble peptides will remain in solution and be recovered in the supernatant while undigested proteins which are not soluble will be found in the precipitate. The supernatant contains a peptide mixture of various amino acid residues and is called the *protein hydrolysate*. The supernatant can be freeze-dried into a powder product and stored frozen for subsequent analysis for bioactive properties, or the solution can be used "as is" for further processing such as in ultrafiltration (separation according to peptide size). Ultrafiltration can be carried out using a series of membranes with designated molecular weight cutoffs. Depending on the starting raw material, the protein hydrolysate could contain various levels of nonprotein products such as sugars, soluble fiber, hydrophilic lipid compounds, and especially salt (NaCl). The content of NaCl is dependent upon whether acid neutralization of the digest is carried out at the end of the reaction.

In order to inactivate the hydrolytic enzyme, the reaction mixture pH can be changed to a value that is outside the activity range of the enzyme. For example, if an enzyme works in the pH 6–9 range, then HCl can be added to reduce pH to an acidic value to denature and inactivate the enzyme protein. However, this process of adding acid can lead to excess production of NaCl in the hydrolysate. A better alternative to acid neutralization is to place the reaction mixture in an environment (water bath or heating block) with temperatures above 80°C and hold for 10–20 min, which will also stop enzyme reaction but without the negative side effect of NaCl production. In general, the level of nonpeptide compounds can be minimized in the protein hydrolysate by using high protein materials such as protein isolates instead of defatted flour. The protein hydrolysate may be used "as is" to test for bioactive properties, or it can be subjected to further fractionation protocols to obtain active fractions or to achieve peptide purity. Pure peptide fractions can be collected and analyzed for amino acid sequence information, which is required to evaluate structure-function properties of bioactive peptides. A generalized schematic representation of typical protocols used in preparation of bioactive protein hydrolysates and peptides is shown in Fig. 3.2.

The extent to which a protein is hydrolyzed by a protease is normally expressed as the degree of hydrolysis (DH), which represents the percentage of peptide bonds cleaved in reference to the total number of peptide bonds in the native protein. DH is inversely related to peptide chain length with high DH values being desirable for the production of very short-chain peptides (2–7 amino acid residues) that have high potential to be absorbed intact from the GIT. DH can vary depending on the type of enzyme used, but usually, for single enzyme-dependent protein hydrolysis, DH can be as high as 30–50% depending on the method used for calculation. When two or more enzymes are used, especially those that contains both endo- and exoprotease activities, higher DH of up to 90% for the resultant protein hydrolysate is possible. However, it should be noted that very high DH may produce bitter protein hydrolysates depending on the type of

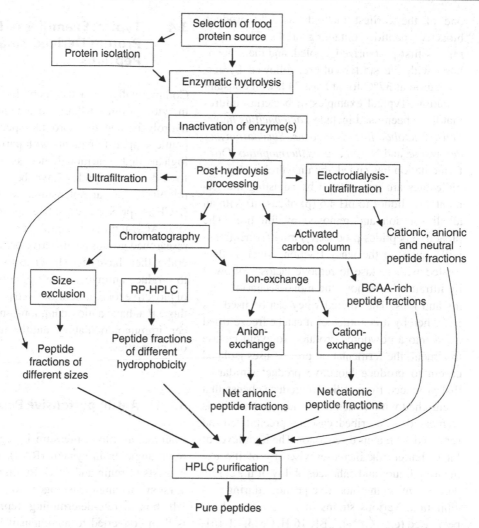

Fig. 3.2 Schematic flow diagram showing various steps involved in the production and processing of food protein-derived peptides (Reproduced from Udenigwe and Aluko (2012) with permission of John Wiley & Sons, Inc)

enzyme and protein starting material. Bitterness of protein hydrolysate is due to the high levels of low molecular weight peptides that contain hydrophobic amino acids. The choice of protease can affect bitterness of the resultant protein hydrolysate, e.g., alcalase generally produces more bitter peptides than papain or chymotrypsin. However, bitterness of protein hydrolysate can be reduced using adsorption chromatography whereby the column is packed with activated carbon or the protein hydrolysate can be treated with exoproteases. The bitterness is reduced because hydrophobic bitter peptides will adsorb to the

activated column while less hydrophobic (less bitter) peptides are obtained as the flow-through. It is not recommended to subject a protein hydrolysate to bitterness removal if bioactive peptides that are responsible for potency of the protein hydrolysate are dominated by hydrophobic peptides.

3.3.2 Microbial Fermentation System

Bacteria fermentation of fat-free milk using predominantly lactic acid-producing cultures was one of the earliest methods used to produce bioactive peptide-containing drinks. The liquid milk is first pasteurized, cooled, and then inoculated with the starter culture, followed by fermentation at 37°C for at least 24 h with constant agitation. Typical examples of bacteria cultures that have been used include *Lactobacillus helveticus*, *Lactobacillus rhamnosus*, *Saccharomyces cerevisiae*, and *Streptococcus thermophilus*. Once fermentation has ended, the unhydrolyzed casein molecules are precipitated by adjusting the fermentation liquor to pH 4.6 (pI of casein) with an alkali solution and removed by filtration. The bioactive peptides produced during fermentation will remain in the whey fraction, which is then treated with lactase to remove lactose followed by filtration to remove the monosaccharide sugars and ions. The filtered whey can be used "as is," whereby it is bottled or it can be freeze-dried dried into a powdered product. To avoid lactase treatment, the fermentation process uses isolated casein to produce bioactive product similar to that produced from milk. To produce cheese that is enriched with bioactive peptides, buttermilk is fermented as described; casein is precipitated and removed with a curd separator. The high level of fat in buttermilk increases viscosity of the fermenting liquor and enhances ability of the casein curd to entrap the bioactive peptides during precipitation. Various strains of *L. helveticus* have been used (e.g., CPN4, LBK 16-H, CM4, LB161, and LB230), but there is almost no differences in the types of peptides produced. A combination of different bacteria strains also seem to be no more effective than using just one strain.

Soybean fermentation has been carried out using *Aspergillus egypticus*. Prior to fermentation, the soybean seeds are soaked in water at 30°C for 4 h followed by steaming (sterilization) at 121°C. After inoculation with *A. egypticus* spores, primary fermentation was carried out at 30°C and 90% humidity for 48 h. This was followed by a secondary fermentation in sealed bottles at 37°C for 15 days after which the product was ready to be consumed as food. Bioactive property of the fermented soybean food product was dependent on length of fermentation with greatest activity obtained after 15 days.

3.4 Typical Examples of Food Protein-Derived Bioactive Peptides

Commercially, bioactive peptides are produced in vitro from customized enzymatic protein hydrolysis, and they provide specialized therapeutic support to patients with particular physiological and nutritional needs. For example, food-derived peptides have been produced as inhibitors of angiotensin-converting enzyme (ACE), a principal causative agent of hypertension. ACE-inhibitory peptides are one of the most researched groups of bioactive peptides, and several of them have been shown to also work in vivo to produce antihypertensive effects. Other types of bioactive food protein-derived peptides include those that have antioxidant, ion-binding, anticancer, immunomodulating, antithrombic, and opioid properties.

3.4.1 Antihypertensive Peptides

Mammalian blood pressure is regulated by the renin-angiotensin system (RAS), which mainly consists of renin and ACE. Renin catalyzes conversion of angiotensinogen to angiotensin I, which is the rate-determining step. Angiotensin I is then converted to angiotensin II (Ang II) by ACE. Ang II is a potent vasoconstrictor that maintains normal contractions of blood vessels at optimum rates under physiological conditions. However, a disorder of the RAS can lead to excess levels of ACE and thus excessive constriction of blood vessels, which if left uncontrolled leads to hypertension and associated cardiovascular problems such as kidney malfunction and left ventricular hypertrophy. Therefore, ACE and renin inhibitors act to reduce the ultimate production of vascular Ang II in hypertensive patients and help to reduce or regularize blood pressure. Aliskiren is the only pharmacological agent approved as renin inhibitor while several drugs (captopril, enalapril, lisinopril, etc.) are available as effective ACE inhibitors. However, due to negative side effects such as dry cough and edema that are

Fig. 3.3 The blood pressure-regulating renin-angiotensin system pathway showing molecular targets (renin and angiotensin-converting enzyme) for bioactive peptides (Reproduced from Udenigwe and Aluko (2012) with permission of John Wiley & Sons, Inc)

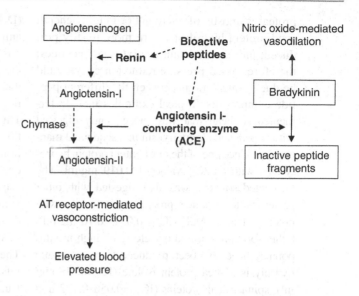

associated with use of antihypertensive drugs, there have been increased efforts aimed at developing food protein-derived peptides as suitable alternatives. Food peptides have been shown to produce no negative side effects after prolonged use and are considered safer than drugs for the treatment of hypertension. However, most often, the effective blood pressure-reducing dosage of peptides is much higher than that of drugs; therefore, peptide products are usually more expensive than most of the generic antihypertensive drugs. The first set of food protein-derived ACE-inhibitory peptides was reported in 1979 by Japanese scientists who hydrolyzed gelatin with collagenase obtained from *Clostridium histolyricum*. This was followed by identification of the famous milk tripeptides (VPP and IPP) that are responsible for the antihypertensive properties of certain fermented milk products. Since then, there has been a substantial increase in the number of ACE-inhibitory peptides (currently over 300) purified from various food sources. Apart from purified peptides, several food protein hydrolysates have been shown to be effective in vitro and in vivo as ACE inhibitors and as antihypertensive agents. In fact, a few products have been commercialized, especially in Japan and several European countries. The RAS and reactions that are known to be susceptible to peptide inhibition are shown in Fig. 3.3.

Potency of protein hydrolysates and purified peptides is expressed as the concentration that inhibits 50% of ACE activity and is referred to as IC_{50}; the lower the value, the more potent the protein hydrolysate or peptide. As indicated above, food protein-derived ACE-inhibitory peptides are not as potent as drugs in terms of concentrations that are required to cause blood pressure reduction or to inhibit ACE activity during in vitro tests. While the therapeutic efficiency (IC_{50} values) of most drugs is at the nanomolar (nM) level, most food protein-derived ACE-inhibitory peptides are only effective at micromolar (μM) or millimolar (mM) concentrations. However, food peptides have the advantage that they are very unlikely to cause intoxication when consumed in large doses that exceed levels required for therapeutic activity, since the body recognizes the excess as normal proteins and processes them accordingly. In contrast, excessive dose of drugs could cause toxic side effects in the body. So far, several human intervention trials have showed no adverse side effects during the use of food peptides as therapeutic agents.

3.4.1.1 ACE-Inhibitory Properties and Antihypertensive Effects of Food Protein Hydrolysates

Several ACE-inhibitory food protein hydrolysates have been produced using different types of proteases, mostly of microbial origin. In general, food protein hydrolysates are more readily produced, and they contain a mixture of peptides; therefore, blood pressure reduction effect could be due to interactions between the different peptide components. Typical examples include fermented soybean that was subsequently digested with pepsin and chymotrypsin in a sequential manner to produce one of the most potent protein hydrolysates with an IC_{50} value of 0.019 mg/ml. The fermented soybean was also digested with other proteases to produce protein hydrolysates with potency against ACE (IC_{50}=0.051−0.16 mg/ml). Other food protein hydrolysates with high in vitro potency have also been produced from enzymatic hydrolysis of pea protein isolate (IC_{50}=0.07 mg/ml), spinach leaf proteins (IC_{50}=0.056–0.172 mg/ml), quinoa seed proteins (IC_{50}=0.075 mg/ml), alfalfa proteins (IC_{50}=0.088 mg/ml), sesame seed proteins (IC_{50}=0.063 mg/ml), wheat gluten (IC_{50}=0.034 mg/ml), and seaweed (IC_{50}=0.099 mg/ml). The potency of a protein hydrolysate is influenced by the type of substrate protein and also by the type of enzyme used for digestion. This is because amino acid sequence differs between various proteins, and proteolytic specificity of enzymes also differs; therefore, hydrolytic products will vary according to enzyme specificity and protein primary structure. For example, hydrolysis of wheat gluten by actinase, chymotrypsin, papain, trypsin, and protease M produces protein hydrolysates with different IC_{50} values of 0.034, 0.418, 0.039, 0.312, and 1.0 mg/ml, respectively. The main reason is because the point of cleavage by an enzyme is dependent on the type of amino acid involved in the peptide bond formation. Therefore, different enzymes will produce peptides with different amino acid patterns. Because amino acid sequence of peptides is considered to be a critical determinant of ACE-inhibitory activity, it means that the inhibitory activity of protein hydrolysates produced from same native protein but with different enzymes may not always be the same (Table 3.1).

Apart from the demonstrated in vitro potency of these plant protein hydrolysates against ACE, several works have also shown that some of the products are capable of reducing systolic blood pressure (SBP) and diastolic blood pressure

(DBP) when they are orally fed to experimental animals. This is very important because the observed blood pressure-lowering effects show the potential for use of protein hydrolysates as therapeutic agents, especially through formulation into functional foods and nutraceuticals. Thus, bioactive protein hydrolysates contain peptides that are able to resist gastrointestinal digestion, be absorbed in intact form into the blood circulatory system, and reach target organs or tissues. Several animal feeding and human intervention trials have so far been conducted to demonstrate the blood pressure-lowering effects of various food protein hydrolysates (Table 3.2). The stroke-resistant spontaneously hypertensive rats (SHR) as well as diet-induced (usually high fructose or high lipid or high salt) hypertensive rats have been frequently used in evaluating hypotensive effects of food protein hydrolysates. SHR have been traditionally bred to develop hypertension starting from the age of 10 weeks because of high level of activity of RAS, which leads to excessive amounts of Ang II in the blood vessels. Males exhibit average systolic blood pressures greater than 200 mmHg by 3–4 months of age, which mimics the level of blood pressure in hypertensive human beings. Various protein hydrolysates obtained from wheat gluten, milk proteins (casein and whey), buckwheat, soybean sesame seed, rapeseed, alfalfa, and spinach leaf have been tested in animal models to determine blood pressure-lowering effects. Decreases in systolic blood pressure of up to 40 mmHg have been reported for some food protein hydrolysates. Oral administration (2 g/kg body weight) of protein hydrolysate from whole soybean seed produced a weaker hypotensive effect (decrease in SBP of 10 mmHg) in the SHRs. Higher antihypertensive potency in SHR has been reported for other protein hydrolysates with concentrations of 10–200 mg/kg body weight being shown to reduce blood pressure. Apart from the oral route, intraperitoneal administration of 1 g/kg body weight of corn gluten hydrolysate was shown to produce a persistent hypotensive effect that lasted up to 6 h and resulted in a decrease of SBP by up to 75 mmHg. Hypotensive effects of food protein hydrolysates usually last from 2 to 8 h depending

Table 3.1 Inhibition of angiotensin-converting enzyme by food protein hydrolysates. Activity is dependent on both the native substrate and type of enzyme used for proteolysis

Native protein or substrate	Proteolytic enzyme	In vitro activity (IC_{50}, mg/ml)
Skate fish	Alcalase	1.890
Skate fish	Neutrase	3.550
Skate fish	Pepsin	6.190
Skate fish	Trypsin	1.130
Alfalfa protein concentrate	Devolase	0.088
Pea protein isolate	Pepsin+chymotrypsin	0.070
Sweet potato protein	Protease S	0.156
Sweet potato protein	Protease M	0.658
Sweet potato protein	Nucleisin	0.376
Sweet potato protein	Thermoase PC 10 F	0.177
Sweet potato protein	Protease N	0.505
Fermented soybean	Pepsin	0.160
Fermented soybean	Trypsin	0.610
Fermented soybean	Pepsin + chymotrypsin	0.019
Fermented soybean	Pepsin + trypsin	0.057
Soybean protein isolate	Pepsin + pancreatin	0.280
Soybean protein isolate	Protease D3	0.180
Defatted soybean meal	Alcalase	0.340
Squid gelatin	Protamex	1.060
Squid gelatin	Alcalase	0.340
Squid gelatin	Neutrase	0.630
Wheat gluten	Papain	0.039
Wheat gluten	Trypsin	0.312
Wheat gluten	Chymotrypsin	0.418
Wheat gluten	Protease M	1.000
Spinach rubisco	Pepsin+pancreatin	0.072
Spinach rubisco	Pancreatin	0.172
Spinach rubisco	Pepsin	0.064
Sea squirt (*Styela clava*)	Papain	2.282
Sea squirt (*Styela clava*)	Protamex	1.023
Sea squirt (*Styela clava*)	Alcalase	1.781
Sea squirt (*Styela clava*)	Trypsin	2.427
Sea squirt (*Styela clava*)	Chymotrypsin	2.263
Sea squirt (*Styela clava*)	Flavourzyme	2.343
Muscle myofibril	Chymotrypsin	0.710
Muscle myofibril	Papain	0.370
Muscle myofibril	Trypsin	0.460
Sarcoplasmic proteins	Papain	0.530
Sarcoplasmic proteins	Chymotrypsin	0.940
Sarcoplasmic proteins	Trypsin	0.720

on the experiment, and long-term feeding for up to 12 weeks has also been shown to be effective. There is evidence that reduction in blood pressure is due to actual in vivo enzyme inhibition as shown by the fact that there were significantly lower levels of ACE activities in plasma, lungs, testes, aorta, and kidneys of SHRs that were fed with protein hydrolysates when compared to the

Table 3.2 Systolic blood pressure-reducing effects of some angiotensin-converting enzyme-inhibitory food protein hydrolysates as measured using spontaneously hypertensive rats

Type of protein hydrolysate	Dose (mg/kg body weight)	Outcome (mmHg) (maximum decrease)
Buckwheat	100	−29
Spinach leaf	250 and 500	−11
Rapeseed	150	−15
Soybean	100–1,000	−38
Ovalbumin	20	−12
Wheat gluten	500	−40
Sweet potato	500	−33
Pea protein	100–200	−18
Sesame seed	1 and 10	−25
Alfalfa	500	−30
Bovine whey	8	−55
Corn	100	−34
Soft-shelled turtle	1,000	−25
Hempseed	200	−32

control animals that received only saline. Renin activity may also be attenuated to cause blood pressure reduction because reduced kidney expression of renin mRNA has been reported in rats that consumed pea protein hydrolysate. The hypotensive potency of protein hydrolysates from the same type of parent protein, e.g., soybean, may not be comparable if produced with different enzymes. This is because of differences in type of peptides produced by each enzyme in addition to variations in hydrolysis conditions, all of which can lead to differences in the ratio of active peptides present in the protein hydrolysates. Overall, there is consistent data to show that repeated administration of food protein hydrolysates could be used as a therapeutic tool to manage arterial hypertension.

Results from human intervention trials have involved mostly milk protein hydrolysates, either the fermented product or the product produced from in vitro protease digestion, but hypotensive effects of pea protein hydrolysate have also been demonstrated in humans. In human trials, modest depressions in blood pressure have been commonly achieved when compared to the stronger blood pressure-lowering effects reported from animal experiments. Fermented milk has been shown to lower SBP by up to 7 mmHg and DBP by about 4 mmHg in hypertensive human subjects. The hypotensive effects of the fermented milk product have been attributed mostly to the presence of the originally discovered ACE-inhibitory tripeptides (IPP and VPP), though contributions from the high levels of calcium, potassium, and magnesium cannot be totally excluded. The tripeptides have been shown to be bioavailable since they can be detected in the aorta of hypertensive rats following oral administration of fermented milk. The contributions of other fermented milk ingredients (apart from the tripeptides) to blood pressure reduction is very likely because there have been nonsignificant changes in plasma ACE activity.

3.4.1.2 Purified Peptides as ACE Inhibitors

Protein hydrolysates consist of peptide mixtures that can differ both in size and in their ACE-inhibitory activities in addition to completely inactive peptides. In order to increase potency of the peptides, purification is usually carried out using bioassay-guided fractionation to separate the peptides into more homogeneous groups or even obtain individually purified peptides. Peptide purification is usually carried out using packed column methods involving gel permeation (GPC) or high-pressure liquid chromatography (HPLC). Most commonly, both methods are used in sequence with GPC first and then preparative HPLC second or vice versa. During HPLC protocols, a reverse-phase column (hydrophobic packing, usually a C18 or C12) is preferred for peptide purification. Several chromatographic runs are needed to be carried out in order to obtain enough peptide materials for subsequent analyses. Once purified, the peptides can be analyzed for amino acid sequence, which enables determination of the relationships between amino acid positional arrangement and ACE-inhibitory potency. Purity can be confirmed using an analytical HPLC, whereby a single narrow peak indicates presence of mostly one peptide specie. Mass spectrometry showing predominance (>95%) of one peptide mass can also be used to confirm purity of the peptide preparation. For short-chain peptides, amino acid sequence can be obtained from the daughter ions (from amino acid fragments) generated during analysis by tandem mass spectrometry. The peptide can also be passed through an automated sequencer, which employs the Edman degradation method to identify each amino acid as the peptide is fragmented.

Because of the relatively low yield of purified peptides, knowledge of the amino acid sequence allows chemical synthesis of the peptide on a larger scale, which could be more cost-effective than the column separation of enzymatic protein digest. Table 3.3 provides a summary of some of the known sequences of ACE-inhibitory peptides that have been shown to reduce blood pressure during in vivo experiments that used SHR. It is a well-known fact that the potential blood pressure-reducing effects of purified peptides cannot be truly assessed using only the in vitro inhibition of ACE activity. This is because of the potential for digestive enzyme-induced destruction and/or inactivation of peptides during passage through the GIT. ACE-inhibitory peptides have been classified according to their metabolic fate in the presence of ACE or GIT enzymes. The first group is called *substrates*, which are peptides that show tentative ACE inhibition during in vitro screening; however, they are inactive and have no effect on blood pressure when given orally to experimental animals. The *substrates* are usually susceptible to degradation when incubated with ACE, which is confirmed as differences in the HPLC elution patterns (or number of peaks) in the presence and absence of ACE. For example, LRIPVA has been shown to be a typical *substrate*-type peptide because even though it has an in vitro IC$_{50}$ value of 0.5 μM against ACE that suggests high potency, it does not have any effect on blood pressure when given orally to SHR even at high dosage. Therefore, the structural degradation products are inactive with respect to blood pressure reduction. The second group is the *true inhibitors*, which are peptides that show ACE inhibition during in vitro screening in addition to blood pressure reduction when fed orally to animals. The true inhibitors are resistant to structural degradation when incubated with ACE, and the HPLC elution pattern is similar in the presence and absence of the enzyme. The third group is called the *pro-drug*, which are peptides that show increased blood pressure-reducing effects after they have been incubated with ACE during in vitro or in vivo screening. Thus, pro-drugs are also susceptible to structural degradation in the GIT or by ACE, but one of the products released has higher inhibitory property than the original peptide. Therefore, in order for an ACE-inhibitory peptide to be useful as a therapeutic agent, it must be either a *true inhibitor* or a *pro-drug* group, i.e., the peptide must actually reduce blood pressure when administered orally. The peptides shown in Table 3.3 can be classified as *true inhibitor* and/or *pro-drug* and may be useful agents for the treatment or prevention of human hypertension. For example, peptides such as MRWRD, IAYKPAG, and VWIS can be classified as *pro-drugs* because incubation with ACE (removes the two amino acid residues from the C-terminal) leads to formation of MRW, IAYKP, and VW, respectively, that also show hypotensive properties. In contrast, all the di- and tripeptides are *true inhibitors* of ACE, probably because the short-chain nature confers some degree of resistance to proteolytic enzymes. Apart from decrease in blood pressure, ACE activity in the aorta of SHRs was lower following oral administration of HHL, which shows that the hypotensive effect of the tripeptide is mostly due to enzyme inhibition. Overall, peptide-induced in vitro inhibition of ACE has been well documented, but the actual mechanism of blood pressure reduction has not been fully elucidated. In addition to in vivo inhibition of ACE activity such as demonstrated for HHL, food protein-derived peptides may also reduce activity or gene expression of other enzymes such as renin and chymase that are involved with blood pressure control in mammals. Peptides may also work either by directly being catalyzed to produce nitric oxide (if arginine is present), a vasodilatory agent or they could activate cell signaling pathways that lead to increased generation of nitric oxide (catalyzed by nitric oxide synthases). For example, the ovalbumin-derived hypotensive peptide RADHPF is believed to upregulate nitric oxide

Table 3.3 Systolic blood pressure-reducing effects of some purified food protein-derived peptides using spontaneously hypertensive rats

Peptide sequence	Native protein	Oral dose (mg/kg body weight)	Outcome (mmHg) (maximum decrease)
RIY	Rapeseed	7.5	−11
IY	Rapeseed	7.5	−10
VWIS	Rapeseed	12.5	−13
VW	Rapeseed	7.5	−11
IPP	Bovine milk	1.8	−28
VPP	Bovine milk	3.6	−24
RPLKPW	Soybean	10.0	−22
RPLKPW	Soybean	2.5	−19
RPLKPW	Soybean	1.0	−16
IAYKPAG	Spinach rubisco	30.0	−15
IAYKP	Spinach rubisco	80.0	−12
MRWRD	Spinach rubisco	30.0	−14
MRW	Spinach rubisco	20.0	−20
IAP	Wheat gliadins	150.0	−60
LHP	Marine shrimp	6.0	−35
VY	*Undaria pinnatifida*	1.0	−22
YH	*Undaria pinnatifida*	10.0	−34
KY	*Undaria pinnatifida*	10.0	−26
IY	*Undaria pinnatifida*	10.0	−25
FY	*Undaria pinnatifida*	10.0	−34
IW	*Undaria pinnatifida*	1.0	−14
KVLPVPQ	Casein	1.0	−24
LKPNM	Dried bonito (fish)	8.0	−23
LKP	Dried bonito (fish)	30.0	−50
LRP	Corn zein	30.0	−15
RADHPF	Ovalbumin	20.0	−13
GKKVLQ	Porcine hemoglobin	50.0	−30
FQKVVA	Porcine hemoglobin	50.0	−30
IKW	Chicken muscle	60.0	−17
GEP	Mushroom	1.0	−36
TQVY	Rice	30.0	−40
AY	Corn	50.0	−10

production to cause vasorelaxation effects in hypertensive rats. Similarly, peptides could compete with angiotensin II for binding to vascular angiotensin receptors. Unlike angiotensin II, binding of the food peptide will not activate vascular contraction; therefore, the competitive exclusion of angiotensin II from its receptor can enhance blood pressure reduction. Lastly, some peptides may act as hypotensive agents through modulation of the immune system and lead to activation of various signaling pathways that reduce sodium reabsorption in the kidney,

enhance cellular expression of endothelial nitric oxide synthase, and increase enzymatic degradation of angiotensin II.

3.4.1.3 Structure and Function of ACE-Inhibitory Peptides

The presence of proline, tyrosine, or tryptophan at the C-terminal end of peptides has been shown to contribute to very high ACE-inhibitory properties. This is evident in the amino acid sequence of milk tripeptides, IPP and VPP, which were one of the first to be discovered ACE-inhibitory peptides. The presence of amino acids with aromatic side chains has also been shown to potentiate ACE-inhibitory properties of peptides. For example, VLIVP, DLP, LVY, LQP, LKY, VIY, MLPAY, VW, MRW, GQP, IAP, LRW, IKP, and YQY are peptides with high potency against ACE activity, and all of them contain either pro-line, tyrosine, or tryptophan at the C-terminal. Initial work has shown that the presence of a hydrophobic amino acid at the C-terminal is also a very important determinant of the overall binding ability of peptides to the active site of ACE. Such peptides compete with angiotensin I (natural ACE substrate) for the active site of ACE and are known as competitive inhibitors. Other researchers have proposed that amino acids with bulky or aromatic side chains are preferred for dipeptides (e.g., VW). For tripeptides, aromatic amino acid at the C-terminal along with a positively charged residue in middle position and a hydrophobic residue at the N-terminal (LKY, MRW, LRW) are the most preferred for high potency against ACE. The presence of tyrosine, proline, and phenylalanine at the C-terminal or phenylalanine at the penultimate position is believed to contribute to the potency of ACE-inhibitory tetrapeptides. Arginine, histidine, tryptophan, and phenylalanine at the C3 position or valine, isoleucine, and methionine at the C4 position also are important requirements for potency of ACE-inhibitory tetrapeptides. For peptides that are longer than tetrapeptides, the type and arrangement of the last four amino acid residues at the C-terminal are known to determine in vitro potency against ACE. Tyrosine and cysteine at the C-terminal or histidine, tryptophan, and methionine at the penultimate C-terminal position in long-chain (>4 amino acid residues) peptides are important contributors for ACE inhibition. Also for long-chain peptides, isoleucine, leucine, valine, and methionine at the C3 position or tryptophan at the C4 position may contribute to enhancing ACE inhibition. However, work is still needed to determine the best combinations of amino acids at individual positions on the peptide chain that contribute most to ACE-inhibitory properties. Because most ACE-inhibitory peptides are di-, tri-, and tetrapeptides,

there is a high probability that they can be absorbed in intact form from the digestive system and be transported in active form to reach target organs and tissues.

3.4.2 Antilipidemic and Antidiabetic Peptides

The ability of protein hydrolysates to reduce blood glucose level has been demonstrated in animal experiments. Apart from the high blood glucose levels, diabetic patients also suffer from vascular complications such as development of atherosclerotic plaques in blood vessels. Thus, it is very important that therapeutic intervention leads to decreased blood glucose and lipid levels in order to mitigate the damaging effects of diabetes. Various protein hydrolysates have been shown to decrease blood glucose and plasma insulin levels in mice models of diabetes. The structure and biological functions (insulin-releasing activity) of the pancreas were greatly improved by feeding type 2 diabetic mice with silk protein hydrolysate. The active principle in silk protein hydrolysate was traced to the <1 kDa peptide fraction, which suggests that small peptides with high potential for absorption from the gastrointestinal tract are responsible for the observed insulin secretagogue (pancreas stimulation for improved insulin production) effects. In addition to improving glucose tolerance, the silk protein hydrolysate and peptides also attenuated high-fat-diet-induced body weight gain in addition to reduced plasma levels of total cholesterol, LDL cholesterol, and the atherogenic index. The effects of silk protein hydrolysate were observed when orally administered to mice at levels of 100 and 200 mg/kg body weight per day. By increasing leptin expression in cells, these peptides reduce adipogenesis and fat accumulation in the adipose tissue. This is because leptin is a known suppressor of adipogenesis and thus plays a vital role in regulation of body energy. Silk peptides also inhibited adipogenesis of adipocytes through downregulation of adipogenic gene expression and protein synthesis. The antidiabetic effect of peptides can also be attributed to ability to upregulate glucose transporters (GLUT). For example, silk protein hydrolysate significantly increased expression levels of GLUT4 in adipocytes, which enhanced glucose uptake into cells and reduce blood glucose levels. In human patients with non-insulin-dependent diabetes, oral administration of silk protein hydrolysate led to improved insulin-releasing activity coupled with reduction in blood glucose.

Several food proteins have been reported to contain bioactive peptide sequences with capacity to reduce blood total lipids and cholesterol. A 24-amino acid peptide (MW 2.27 kDa) present within the α′ subunit of soybean conglycinin has been shown to possess cholesterol-lowering properties by increasing LDL-receptor-mediated LDL uptake in Hep G2 cells. Another peptide sequence, an octapeptide (FVVNATSN) was isolated from the enzymatic digest of soybean proteins and shown to be a very potent stimulator of LDL-receptor transcription in Hep T9A4 human hepatic cells. A soybean protein hydrolysate with molecular weight <10 kDa also showed potential hypotriglyceridemic properties by altering gene expressions related to triglyceride synthesis and also decreased Apo B-100 accumulation in Hep G2 cells partly as a result of an increase in LDL-receptor mRNA expression. Apo B-100 is the major functional apolipoprotein component of very-low-density lipoproteins (VLDL); degradation of Apo B-100 reduces VLDL synthesis with concomitant decreases in plasma lipid and cholesterol. In addition to modulations in gene expressions, soy protein hydrolysates and constituent peptides can also bind to bile acids and neutral sterols in the intestine to enhance fecal excretion, which produces hypocholesterolemic effects. The bile acid-binding ability of soy protein hydrolysates is partly dependent on the content of insoluble high molecular weight (HMW) peptide fraction that is rich in hydrophobic amino acids. The undigested and insoluble hydrophobic peptides are able to bind to cholesterol and bile acids to prevent reabsorption from the GIT and reduce plasma lipid content. Thus, not all the peptides necessarily need to be absorbed in order to produce cholesterol-lowering effects. A soybean protein-derived hydrophobic peptide (WGAPSL) has been shown as the potent agent

in a hypocholesterolemic protein hydrolysate produced from alcalase digestion of soybean proteins. When tested in hypercholesterolemic mice, the WGAPSL-containing soy protein hydrolysate was able to reduce serum total and VLDL+LDL cholesterol. Another soybean protein-derived hydrophobic hexapeptide peptide (VAWWMY) also has been shown to have bile acid-binding properties. Genetically modified soybean glycinin that contains copies of VAWWMY has an improved bile acid-binding property over the nonmodified protein. The high hypocholesterolemic properties of VAWWMY can be attributed to the presence of mostly hydrophobic amino acids, which enhances interactions with the equally hydrophobic bile acids. It is possible that interaction of the peptide with bile acids reduces or prevents digestion of VAWWMY, which enhances fecal removal of cholesterol and other sterols. Thus, VAWWMY peptide is a potential agent that can be used to formulate nutraceuticals aimed at reducing bile acid reabsorption in the GIT and provide hypocholesterolemic effects and associated cardiovascular health benefits.

3.4.3   Opioid Peptides

These are peptides that possess some form of affinity for opiate receptors and can exert effects on the nervous system. Naturally occurring opioid peptides in the body (usually brain and pituitary gland) are called endorphins and enkephalins while opioid peptides obtained from enzymatic digests of food proteins are called exorphins (because they are of exogenous origin and morphine-like activity). Generally, most of the opioid peptides that are derived from food proteins have N-terminal amino acid sequence of Y-X-F or Y-X1-X2-F whereas typical endogenous opioid peptides have similar amino acid sequence (YGGF) at the N-terminal. Initially, food protein-derived opioid peptides were isolated from dairy proteins (caseins) and were called the β-casomorphins. Subsequently, whey proteins have also been found to contain opioid-like amino acid sequences such as fragment 50–53 (YGLF) in α-lactalbumin and fragment 102–105 (YLLF) in β-lactoglobulin; the two peptides are called α-lactorphin and β-lactorphin, respectively. Therefore, the key structural features include presence of (1) a tyrosine residue at the amino terminal end (except some α-casein opioids) and (2) another aromatic residue, phenylalanine or tyrosine, in the third or fourth position from the N-terminal residue. It is these structural features that enable the peptides to fit properly into the binding site of the opioid receptors while differences in opioid activity of peptides may be related to differences in conformational flexibility. α-Lactorphin can be produced through pepsin-mediated digestion of α-lactalbumin while β-lactorphin is produced by consecutive digestion of β-lactoglobulin with pepsin and trypsin (or simultaneous digestion with trypsin and chymotrypsin). α-Lactorphin and β-lactorphin have been shown in vitro to inhibit and stimulate, respectively, the contractions of guinea pig ileum. Whereas liberation of casomorphins has been shown to occur during gastrointestinal digestion of caseins, there is currently no evidence that lactorphins are released during in vivo digestion of α-lactalbumin and β-lactoglobulin. In general, below are some of the known functions of these opioid peptides:

- Have morphine-like properties and increase analgesic behavior
- Modulate social behavior
- Prolong gastrointestinal transient time by inhibiting intestinal peristalsis and motility
- Stimulate endocrine responses such as secretion of insulin and somatostatin
- Stimulate uptake of high-fat diet
  There are two types of opioid peptides:
(a) Opioid agonists whose characteristics include:
  - Modulation of social behavior
  - Increase of analgesic behavior
  - Prolongation of gastrointestinal transient time by inhibiting intestinal peristalsis and motility
  - Exertion of antisecretory (antidiarrheal) action
  - Modulation of amino acid transport
  - Stimulation of endocrine response such as secretion of insulin and somatostatin
  - Stimulation of high-fat diet intake
(b) Opioid antagonists whose characteristic feature is the suppression of the activity of agonists

3.4.4 Caseinophosphopeptides (CPP)

These are phosphorylated peptide fragments obtained from enzymatic hydrolysis of casein (milk protein), which is known to contain several phosphoserine residues (the phosphate group is attached to serine). At physiological pH, the phosphate groups carry a negative charge, which enhances binding to positively charged cations, especially dietary calcium that has relevance to improving human health. The complex formation provides calcium in the organic form, and absorption is enhanced in the small intestine when compared to the inorganic calcium. Because they provide calcium in a highly bioavailable form, the CPP also inhibit caries lesions by enhancing recalcification of the dental enamel. Hence, it has been proposed that these peptides may be used to bind calcium and provide a dietary product suitable for the treatment of dental diseases. CPP can act as a transport mechanism for cations in the gastrointestinal tract and thereby improve ion solubility, which enhances bioavailability. The phosphoserine-rich amino acid sequences are naturally present in breast to protect the milk gland against calcification by controlling calcium phosphate precipitation. Cleavage of casein with proteolytic enzymes such as trypsin can release CPP fragments. It has been shown that rats fed casein had CPP in the small intestine after approximately 2.5 h. CPP fragments are known to be resistant to proteolytic breakdown in the intestinal tract, which enhances their ability to exert in vivo effects.

3.4.4.1 Summary of CPP Applications

- Drinks fortified with calcium and iron in the presence of CPP, for use in treating osteoporosis and anemia, respectively.
- Products for children that incorporate calcium or other minerals and CPPs in sweets or cookies.
- Other possible uses: infant nutrition, calcium-enriched dairy products, and natural calcium supplements.
- Potential use in functional food confectionery. Complexes of calcium, CPPs, and phosphate may reduce dental caries. This is because the CPP-calcium-phosphate complex increases

Table 3.4 Changes in calmodulin secondary structure fractions in the presence of flaxseed protein-derived inhibitory peptides (Adapted from Omoni and Aluko (2006). Reproduced with permission (John Wiley & Sons))

Treatment	α-Helix (%)	β-Sheet (%)	β-Turn (%)	Random (%)
None	49	13	15	24
Ca^{2+}	62	11	13	16
0.5 mg/ml Fraction I[a]	53	28	4	19
0.5 mg/ml Fraction II[b]	54	21	6	19

[a]First eluted fraction from SP-Sepharose column and contained 42% (w/w) of positively charged amino acid residues
[b]Second eluted fraction from SP-Sepharose column and contained 50% (w/w) of positively charged amino acid residues

level of calcium phosphate in the plaque and can significantly reduce adherence of streptococci to the teeth.

3.4.5 Calmodulin-Binding Peptides

Calmodulin (CaM) is a calcium-activated protein that plays important roles in maintaining physiological functions of cells and body organs including cell division, cell proliferation, and neurotransmission. CaM has a net negative charge at physiological pH and is a 16.7-kDa calcium-binding protein that contains 148 amino acid residues with multifunctional roles in the translation of intracellular messages. Some enzymes that are activated by and have obligatory requirements for CaM include protein kinase II, nitric oxide synthases, and phosphodiesterase I. While CaM functions are vital for cell survival, excessive levels can pose serious problems within the body with subsequent development of chronic diseases such as cancer, cardiac hypertrophy, and Alzheimer's disease. Therefore, bioactive inhibitors such as food protein-derived peptides are designed to bind to CaM in order to prevent or treat associated diseases. The mechanism involved in peptide-induced inhibition of CaM-activated reactions is twofold. One, the peptide inhibitor should have a net positive charge at physiological pH, which enhances binding to the negatively charged CaM. This was confirmed by the fact that acetylation of positively charged peptides led to decreased interactions with CaM and reduced ability of the peptides to modulate activity of

CaM-dependent enzymes. Two, the peptide inhibitor should have some hydrophobic amino acid residues, which enhances binding with CaM through hydrophobic interactions. This is especially important because calcium is required for activation of CaM, and binding to calcium leads to structural changes in the protein that cause increased exposure of hydrophobic patches. In both cases, binding of peptide inhibitors leads to changes in molecular conformation of CaM, which reduces ability to interact with and activate target enzymes. For example, binding of Ca^{2+} to CaM causes an increase in the α-helix structure (Table 3.4), which enhances ability of CaM to activate target enzymes. However, the fraction of α-helix structure becomes reduced when peptide inhibitors bind to CaM, which reduces activation of CaM-dependent enzymes.

CaM-binding peptides can be produced through enzymatic hydrolysis of food proteins and followed by selective fractionation to isolate positively charged peptides as shown in Fig. 3.4.

Hydrolysis of bovine α-casein followed by cation-exchange column fractionation led to identification of peptide sequences that inhibited CaM-dependent phosphodiesterase (CaM-PDE). Table 3.5 shows the amino acid sequences, locations on casein primary structure and CaM-PDE inhibitory activities of the three peptides purified from pepsin hydrolysate of casein. The shortest peptide (16-mer) was the least active while the 17, 19, 24, and 25-mer peptides were the most active. The 16-mer peptide had less number of bulky aromatic amino acid residues (hydrophobic character), which could be responsible for

Fig. 3.4 Production of food protein-derived calmodulin-binding peptides (Reproduced from Aluko (2010) with permission of John Wiley & Sons, Inc)

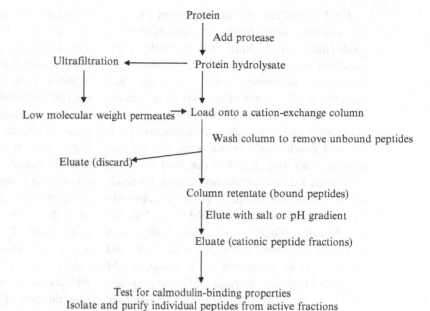

Table 3.5 Inhibition of calmodulin-dependent phosphodiesterase I by peptides isolated from peptic hydrolysate of bovine casein (Adapted from Kizawa et al. (1995))

Peptide sequence	IC_{50} (μM)	Position on the α_{s2}-casein chain
LKKISQRYQKFALPQY	65.0	f164–179
KPWIQPKTKVIPYVRYL	1.1	f191–207
AMKPWIQPKTKVIPYVRYL	3.2	f189–207
VYQHQKAMKPWIQPKTKVIPYVRY	7.0	f183–206
VYQHQKAMKPWIQPKTKVIPYVRYL	2.6	f183–207

the lower CaM-PDE inhibitory property when compared to the 24 and 25-mer peptides that have higher levels of hydrophobic amino acids. Thus, apart from the positively charged amino acid residues, the content of hydrophobic amino acid residues is an important determinant of the inhibitory capacity of food protein-derived CaM-binding peptides.

The physiological importance of CaM-binding peptides has been shown to be related to enzyme inhibition, mainly phosphodiesterase I, nitric oxide synthases (NOS), and protein kinase II. CaM-PDE is an important enzyme that breaks down cyclic adenosine monophosphate (cAMP) within the cell matrix to produce a linear nucleotide phosphate. cAMP is very important in cell homeostasis because it has negative effects on

cell proliferation and cell function through a mechanism that involves irreversible G_1 arrest that is followed by cytolysis. Therefore, cAMP is required for and can induce apoptotic cell death in normal and cancerous cells. However, excessive levels of CaM-PDE causes the depletion of cellular cAMP and prevents physiological regulation of cell death, which can then lead to uncontrolled cell proliferation as seen in cancer pathology. Thus, the inhibition of CaM-PDE activity could elevate cellular levels of cAMP to restore controlled cell death and provide therapeutic relief during cancer. Food protein-derived CaM-binding peptides could serve as suitable ingredients in the formulation of therapeutic products for the prevention and management of cancer development and metastasis.

CaM-dependent NOS are a group of enzymes that are responsible for the production of nitric oxide from arginine. Nitric oxide is a well-known cellular messenger that participates in neurotransmission, vasodilation, immune response, and smooth-muscle contraction-relaxation among several other functions. However, excessive levels of nitric oxide can cause development of pathological conditions such as acute and chronic neurodegenerative disorders that lead to stroke, Alzheimer's disease, Parkinson's, migraine headaches, convulsion, and pain. There are three main isozymes of NOS: neuronal (nNOS), endothelial (eNOS), and inducible (iNOS), each with its own distinct physiological role. Whereas nNOS and eNOS are constitutive enzymes that require CaM binding only prior to activation of target enzymes, iNOS is permanently bound to CaM. Therefore, CaM inhibitors are usually more effective against the constitutive NOS isozymes, since activation of target enzymes could be prevented through prevention of effective interactions between CaM and enzyme proteins. nNOS participates mostly in neurotransmission pathways while eNOS has been implicated in vasodilatory activities and in the production of tumor necrosis factor-alpha (TNF-α) by human monocytes or macrophages. Therefore, eNOS could be regarded as a proinflammatory agent, and inhibitory compounds such as CaM-binding peptides could provide therapeutic benefits during disease conditions.

3.4.6 Antioxidant Peptides

Normal cell metabolism leads to production of reactive oxygen species (ROS) and reactive nitrogen species (RNS), both of which can be highly destructive to cellular components. The body produces antioxidative compounds and enzymes that neutralize these reactive species and prevent oxidative-induced development of chronic diseases. However, due to several factors such as age, genetic predisposition, and environmental conditions, the delicate balance between level of reactive species and natural antioxidant capacity may be tipped in favor of ROS and RNS.

Therefore, exogenous sources of antioxidant capacity could be provided to mitigate the negative effects associated with endogenous antioxidant insufficiency. For food protein-derived peptides to be good sources of physiologically relevant antioxidants, they should have the following two basic structural features. One is the presence of excess or donor electrons and/or aromatic ring that can exist in several stable configurations. The excess electrons can be donated and used to neutralize the free radicals while the aromatic ring ensures that loss of electrons does not change the peptide into another free radical. Therefore, the presence of several amino acid residues in the peptide chain can enhance antioxidative property as a result of the additive effects in terms of electron transfer to the free radicals. Two is the possession of fairly hydrophobic character, which enhances translocation through the bilipid cell membrane structure into the cell cytoplasm and mitochondria where free radicals are generated. Other mechanisms of antioxidant activity of peptides include chelation of transition metals and ferric-reducing power, which contribute to reduced lipid oxidation within the body.

Several food protein hydrolysates and peptides with antioxidant properties in various oxidative reaction systems have been produced from plant and animal protein sources such as pea, soy, fish, quinoa, flaxseed, milk casein, whey, and egg. The demonstrated antioxidant properties of these peptides include scavenging or quenching of ROS/free radicals and inhibition of ROS-induced oxidation of biological macromolecules (lipids, proteins, and DNA). For example, an in vitro study demonstrated the capacity of tuna liver protein hydrolysate to inhibit hydroxyl radical-induced DNA damage on pBR322 plasmid DNA. Addition of tuna protein hydrolysate to DNA prevented the hydroxyl radical-induced conversion of supercoiled plasmid DNA to open circular DNA. Since DNA damage is a known causative factor of various chronic diseases such as cancer, Alzheimer's, and Parkinson's, treatment with antioxidative peptides may provide protection. The mechanism of action involves ability of the tuna protein hydrolysates to scavenge H_2O_2 and hydroxyl radical, the latter being one of the most reactive and destructive free radicals produced within cells. The tuna protein hydrolysates also showed iron-chelating ability, which prevents the Fenton reaction (iron-catalyzed degradation of H_2O_2) as shown in Eqs. 3.1 and 3.2 below.

$$Fe^{2+} + H_2O_2 \rightarrow Fe^{3+} + {}^\bullet OH + OH^- \quad (3.1)$$

$$Fe^{3+} + H_2O_2 \rightarrow Fe^{2+} + {}^\bullet OOH + H^+ \quad (3.2)$$

Therefore, overall effect of the tuna protein hydrolysate (and possibly most bioactive antioxidant peptides) involves prevention of free radical formation through metal chelation and prevention of oxidative damage to the DNA by scavenging of oxygen radicals that may have been formed.

The amino acid composition of food protein hydrolysates have been shown to have strong influence on their antioxidant properties. For example, the amount of histidine, cysteine, proline, methionine, and aromatic amino acids have been reported to contribute to the antioxidant activity of food peptides. Structure-function studies using a number of synthetic peptides revealed that the imidazole group of histidine residue of peptides contributes to increased metal ion chelation, quenching of active oxygen, and scavenging of hydroxyl radical. This is because the imidazole group can participate in hydrogen atom transfer and single electron transfer reactions to neutralize free radicals or bind metal ions. As indicated above, the presence of hydrophobic groups can enhance antioxidant potential as shown by the fact that addition of hydrophobic amino acids, proline and leucine, to the N-terminus of a dipeptide His-His resulted in enhanced antioxidative property of the peptides. Apart from enhancing cellular uptake, hydrophobic amino acids are important for enhancement of the antioxidant properties of peptides since they can increase the accessibility of the antioxidant peptides to hydrophobic polyunsaturated chain of fatty acids within biological membranes to limit oxidative damage. This is important in biological systems because the unsaturated fatty acids in cell membranes are very prone to

oxidative damage by free radicals and oxygen species. Thus, the ability of hydrophobic peptides to interact with cell membranes could provide in situ protection against lipid peroxidation and cellular damage. Moreover, the electron-dense aromatic rings of phenylalanine, tyrosine, and tryptophan residues of peptides can contribute to the chelating of pro-oxidant metal ions whereas phenylalanine can also scavenge hydroxyl radicals to form more stable para-, meta-, or ortho-substituted hydroxylated derivatives. The proton-donating ability of the sulfhydryl (−SH) group in methionine and cysteine has also been shown to enhance antioxidant properties of peptides that contain sulfur-containing amino acids (SCAA). For example, the presence of sulfur-containing and hydrophobic amino acids (but not positively charged amino acids) had a positive effect on the DPPH and H_2O_2 scavenging effects of certain food protein hydrolysates. Acidic amino acids (glutamic and aspartic), which have electron-donating ability, have also been shown to provide strong contributions to the DPPH scavenging, ferric reducing, and H_2O_2 scavenging effects of certain food protein hydrolysates. Likewise, the presence of tryptophan has been shown to contribute to antioxidant properties of peptides, especially hydroxyl scavenging ability. Therefore, the specific contribution of individual amino acid residues to the antioxidant activity of a peptide depends largely on the nature of the ROS/free radical and the reaction medium. However, it is not clear how these "antioxidant" amino acid residues contribute to the antioxidant activity of a peptide mixture typical of food protein hydrolysates or the possible positive or negative contributions of other amino acid residues present in the hydrolysates. This is because mixtures of amino acid that simulate amino acid composition of antioxidant protein hydrolysates have been shown to be not as effective as peptide antioxidants. Therefore, the additive effects of peptide amino acids may be stronger than individual contributions of similar but free amino acids. There is also a need to investigate contributions of amino acid sequence arrangement to antioxidant activities of peptides.

But the major obstacle in determining the structure-function relationships of antioxidant peptides is that several tests (functionality) are involved. Therefore, a peptide could have excellent superoxide and H_2O_2 scavenging effects but poor hydroxyl ion and metal-chelating properties. The solution may be to determine structure-function properties of peptides in relation to individual antioxidant tests. This approach has been used to determine contributions of amino acids to specific antioxidant properties of food protein hydrolysates. Using chemometric analysis of several food protein hydrolysates, it was reported that for superoxide scavenging efficiency of protein hydrolysates, the presence of lysine and leucine residues had strong positive contributions while proline, phenylalanine, and tyrosine had moderate positive contributions. In contrast, the presence of methionine and cysteine had strong negative contributions to the superoxide scavenging efficiency of protein hydrolysates. Ferric-reducing ability of protein hydrolysates was strongly enhanced by the presence of cysteine, methionine, and glutamic acid while lysine had a negative effect. The presence of cysteine, phenylalanine, leucine, isoleucine, proline, and threonine had strong positive effects in enhancing the H_2O_2 scavenging effects of protein hydrolysates, but histidine, lysine, and arginine had negative effects. Ability of food protein hydrolysates to scavenge DPPH radicals was strongly dependent on the presence of aspartic acid, threonine, valine, and isoleucine while histidine, lysine, and arginine had negative contributions. One of the surprising aspects of the chemometric approach is the lack of positive contribution from histidine to any of the four analyzed antioxidant properties. In fact, histidine had negative effects on the radical (DPPH and superoxide) scavenging properties of the food protein hydrolysates. Several reports using simple correlation but comparatively smaller data base (versus the chemometric analysis) have previously indicated strong role for histidine in enhancing different antioxidant properties of food protein hydrolysates. Therefore, further studies are required to confirm the roles of individual amino acids in enhancing antioxidant properties of food protein hydrolysates. However, it should be noted that the previous works with histidine-containing peptides and the chemometric analysis did not involve contributions from amino acid sequence of individual peptides, which may yield different information from the simple amino acid composition data analysis.

In addition to the proton-donating ability of the sulfhydryl functional group, peptides that contain cysteine residues can donate cysteine to be used for the synthesis of glutathione (GSH, γ-L-glutamyl-L-cysteinylglycine), a ubiquitous cellular antioxidant tripeptide. Because GSH is the main nonenzymatic antioxidant compound in cells, it follows that cysteine-containing peptides are useful for enhancing the ability of cells to regenerate their physiological antioxidant defense system. GSH has the protective ability to scavenge free radicals, remove toxic lipid peroxides, repair oxidative damage, and preserve thiol-disulfide status of vital cellular proteins. Moreover, food-derived peptides can also display antioxidant property by induction of gene expression of proteins that protect cellular components from oxidative stress-induced deterioration. In endothelial cells, a dipeptide Met-Tyr derived from sardine muscle protein was shown to stimulate expression of heme oxygenase-1 and ferritin, which led to sustained cellular protection from oxidative stress. Casein hydrolysates produced with different proteases exhibited varying antioxidant activities, independent of degree of hydrolysis, in human Jurkat T cells by increasing cellular catalase activity and amount of reduced GSH. Although the casein hydrolysates showed dose-dependent decrease in viability and growth of the human Jurkat T cells, lower doses retained the beneficial antioxidant properties without any effect on membrane integrity. In D-galactose-induced aging ICR mice, oral intake of jellyfish collagen hydrolysates prepared with Protamex induced increase in superoxide dismutase (SOD) and GSH peroxidase with concomitant decreases in serum and hepatic malondialdehyde (MDA), an oxidative stress marker. While promising results have been obtained from in vitro and animal studies, there is still very scanty information on the ability of food protein-derived peptides to reduce oxidative stress or associated disease symptoms in humans. The following sections provide insight into various food sources of antioxidant peptides and potential mechanisms of action.

3.4.6.1 Hen's Egg Protein-Derived Peptides

The ability of egg yolk phosvitin-derived phosphopeptides to protect Caco-2 cells against H_2O_2-induced oxidative stress was shown to be mediated through suppression of IL-8 secretion, which is a proinflammatory marker. The phosphopeptides containing various levels of phosphorus were obtained through tryptic digestion of phosvitin, the major phosphorylated protein in egg yolk. The phosphopeptides were more effective suppressors of IL-8 secretion than glutathione and the unhydrolyzed but partially dephosphorylated phosvitin. Thus, hydrolysis into smaller oligopeptides had a positive effect on antioxidant properties probably because the reduction in peptide size enhanced uptake into the Caco-2 cells when compared with the larger-sized dephosphorylated phosvitin. In addition to the beneficial effects of smaller peptide size, the levels of phosphorylated residues and histidine also enhanced inhibition of IL-8 secretion. Phosphate groups have excess electrons, and presence of several residues on the same peptide provides a strong source of radical quenching ability. The importance of the additive nature of several phosphate groups on a peptide is apparent because use of phosphoserine (a single amino acid) was not as effective as an antioxidant when compared to the phosphopeptides. The egg yolk phosphopeptides also suppressed MDA production, which is an indication of reduced level of lipid oxidation when compared to untreated control cells. The phosphopeptides also ameliorated the H_2O_2-induced depletion of GSH, which was directly related to increased activity of glutathione reductase (GR), suggesting a protective effect of the peptides against oxidative stress. Under high oxidative stress, the reduced form of GSH is used and become converted to the oxidized form (GSSG). In order to maintain high oxidative defense, the GSSG must be reduced back to two

molecules of reduced GSH, which helps to maintain a desirable high GSH/GSSG ratio. Thus, by stimulating increased activity of GR, the phosphopeptides help the cellular GSH/GSSG recycling system that favors increased availability of reduced GSH and cellular ability to reduce oxidative stress.

3.4.6.2 Myofibrillar Protein Hydrolysates and Peptides

Enzymatic hydrolysates of porcine myofibrillar protein were effective in reducing Fe^{2+}-induced lipid (linoleic acid) oxidation. Antioxidant activity of the hydrolysates was higher at pH 7.1 than at pH 5.1, which may be due to increased ionization of amino acid side chains with concomitant higher number of excess electrons. The ability of the hydrolysates to chelate positively charged metal ions was higher with peptides present in the acidic fraction, which suggests the presence of negatively charged (ionized) groups on the peptides. Peptide hydrophobicity coupled with ionization of carboxyl groups at high pH was more effective in producing strong antioxidant effects than either factor alone. For example, when tested at pH 6–7, myofibrillar protein hydrolysates that contained higher number of hydrophobic amino acids were more effective suppressors of hydroperoxide formation than the hydrolysate with less hydrophobicity. Thus, the hydrophobic character enhanced protective interaction with lipids while acidic groups provided metal ion-chelating ability needed to prevent lipid oxidation. When constituent peptides of the acidic fraction were identified, the sequences were dominated by acidic amino acids (glutamic and aspartic acids). Antioxidant peptides identified from the myofibrillar protein hydrolysates include DAQEKLE (most active), EELDNALN, VPSIDDQEELM, IEAEGE, and DSGVT.

3.4.6.3 Sardinelle Protein Hydrolysates and Peptides

Hydrolysis of sardinelle proteins with various enzymes showed variations in antioxidant properties of the hydrolysates, which were dependent on enzyme type. The antioxidant capacity of the protein hydrolysates increased at higher degree of hydrolysis, suggesting short-chain peptides are better antioxidants than the native long-chain peptides. The improved antioxidant capacity at high DH is consistent with the fact that increased hydrolysis will lead to the availability of more free electrons through increased level of carboxyl groups. Several antioxidant peptides were identified from the sardinelle protein hydrolysates and include LAWL, GGE, LHY, GAH, GAWA, PHYL, and GALAAH. Notable features of the sardinelle antioxidant peptides are the presence of known antioxidant amino acids such as tryptophan (W), histidine (H), and glutamic acid as well as hydrophobic amino acids, leucine (L), alanine (A), and proline (P). LHY with two bulky amino acids and a hydrophobic residue had the highest DPPH scavenging activity, which indicates importance of the imidazole group (binds to lipid radicals and metal ions), tyrosine (electron donor), and hydrophobicity of leucine (enhances interaction between peptides and DPPH) in potentiating strong antioxidant properties of peptides.

3.4.6.4 Egg Ovotransferrin-Derived Peptide

Using cell culture techniques, the potential anti-inflammatory effects of IRW, an egg ovotransferrin-derived peptide, was shown to involve attenuation of TNF-α-induced increase in vascular adhesion molecules. It is a well-established fact that vascular inflammatory response lead to upregulated production of various molecules such as intercellular adhesion molecule 1 (ICAM-1), monocyte chemoattractant protein 1 (MCP-1), and vascular cell adhesion molecule 1 (VCAM-1). Increased levels of these adhesion molecules lead to increased accumulation of oxidized cholesterol as foam cells in vascular walls and contribute to the initiation and progression of cardiovascular diseases. Therefore, by reducing inflammatory response, antioxidant peptides could find potential use in the prevention and management of cardiovascular diseases. The cell culture data showed that the presence of IRW had a dose-dependent effect in reducing the production of MCP-1, ICAM-1, and VCAM-1 by TNF-α-treated human umbilical vein endothelial cells (HUVECs).

TNF-α is a major proinflammatory molecule because it activates production of nuclear factor κB (NF-κB), a molecule that mediates key steps in the pathogenesis of atherosclerosis. The mechanism of action in the anti-inflammatory action of IRW seems to involve prevention of NF-κB-mediated translocation of p65 proteins from the cytosol into the nucleus but not the expression of TNF-α receptors or the degradation of cellular inhibitors of NF-κB. In the inactive state, NF-κB is bound to inhibitory proteins (IκBα and IκBβ) within the cytosol, which prevents NF-κB-mediated translocation of p65 proteins. However, in the presence of proinflammatory factors such as free radicals and TNF-α, the IκB proteins become phosphorylated by IκB protein kinase and are dissociated from NF-κB. The p65 protein is known to enter the nucleus and enhance the expression of proinflammatory proteins. Pretreatment of HUVECs with IRW prevented the NF-κB-mediated nuclear translocation of p65 proteins, which indicates potential role for use of this peptide in managing oxidative stress-related cardiovascular dysfunctions. The IRW peptide also acted as an antioxidant by reducing production of superoxide radicals (a promoter of oxidative stress) in TNF-α-treated or nontreated HUVECs. Therefore, in addition to the attenuated signaling pathways that reduced level of adhesion molecules, the IRW peptide also acted directly within the HUVECs as a scavenger of free radicals to reduce oxidative stress and associated negative effects such as signal transduction. It should be noted that the superoxide radical could serve as signal transduction messenger molecule that induce molecular expression of several proinflammatory factors.

3.4.7 Anticancer and Immune-Modulating Peptides

Immunomodulatory peptides can have effects on both the immune system and cell proliferation response. For example, peptides derived from casein hydrolysates were shown to increase phagocytotic activity of human macrophages against aging red blood cells. Soybean lunasin and peptides from lunasin digestion have been shown to have strong anticancer effects through inhibition of histone acetylation while their anti-inflammatory effects were demonstrated through suppression of NF-κB.

In macrophages, inflammatory response is marked by an increased production of NO and proinflammatory cytokines. Treatment of macrophages with pea protein hydrolysate (PPH) significantly reduced lipopolysaccharide (LPS)/interferon-γ (IFN-γ)-induced NO production in addition to significant suppression of the proinflammatory cytokines, TNF-α, and interleukin-6 (IL-6) production. LPS, an endotoxin present in the outer membrane of Gram-negative bacteria, stimulates phagocytic cells to produce NO and proinflammatory cytokines such as TNF-α and IL-6. High levels of TNF-α and IL-6 have been linked to the pathogenesis of inflammatory diseases and cancer. The unhydrolyzed pea protein had no effects on NO production; therefore, peptides present in the PPH were presumed to be responsible for the observed effects. While physiological concentration is desirable, the production of high levels of NO by macrophages can be cytotoxic and is believed to induce cancer. Therefore, the suppression effect of PPH on NO production by macrophages may be indirectly beneficial in preventing or reducing the incidence of inflammation and cancer. Because carcinogenesis and tumor progression/metastasis can result from persistent state of inflammation that includes recruitment of macrophages, production of ROS and proliferation of genomically unstable cells, peptides that show antioxidant as well as anti-inflammatory properties are considered to be potentially good anticancer agents. The PPH was also nontoxic because the cells remained viable even at high PPH concentration of 25 μg/ml. Oral administration of the PPH (100 μg/day) to BALB/c female mice led to enhanced phagocytic activity of peritoneal macrophages, which is an indication of stimulation of the innate immune system. This is important because macrophages are at the forefront of body's defense during immune response to foreign matter and the increased phagocytic activity indicates higher capacity to eliminate toxic compounds. Moreover, stimulation of the innate immune system in the gut is associated with antitumor activities in mice inoculated with breast cancer cells. The numbers of IgA+cells, IL10+, IL4+, and IFN-γ+cells were also significantly increased in mice that received oral doses of PPH, which suggests enhanced immune surveillance. There was no effect on the number of intestinal IgG+cells, which shows that the PPH did not induce inflammatory immune response in the mice. Similar effects on mice immune response and cytokine production have also been reported for a fermented fish protein concentrate that was enriched with peptides.

Administration of fermented milk (contains bioactive peptides) to mice was shown to suppress breast tumor volume, which was accompanied by increased cellular apoptosis and number of IL-10(+) cells in the tumors but decreased levels of IL-6(+) and TNF-α(+) cells. IL-10 has also been implicated in the immune response associated with colon cancer growth inhibition by yogurt. IL-6 is present in high levels in breast tumors and is a known promoter of cancer growth because it has proangiogenic effects, which can enhance cell proliferation. Activity of TNF-α is related to estrogen synthesis, which at high levels can induce formation of breast cancer or other estrogen-dependent tumors. Therefore, downregulation of TNF-α is an important mechanism by which the anti-inflammatory peptides in the fermented milk can reduce breast cancer formation and disease progression.

The anticancer effects of a rapeseed protein hydrolysate (RPH) have also been demonstrated in a murine model of cancer. Mice transplanted with murine sarcoma S180 cells developed an average of ~1.5 g of tumor weight after 10 days; treatment with the RPH led to dose-dependent 34–53% decreases in tumor weight. RPH-treated mice also had significantly higher thymus index values (produces the T lymphocyte cells), which suggests increased capacity of the immune system to neutralize foreign or toxic substances. The potential effect of RPH in increasing antigen-specific T-cell recruitment was observed in the fact that induction of delayed type of hypersensitivity (DTH) was enhanced at the highest dose

(150 mg/kg body weight/day). Enhanced DTH represents an increased resistance to infection and an improved host defense capability. RPH treatment produced normalized mouse liver pathology and reduced liver pathology as evident in the reduced size of nuclei (enlarged in the untreated mice). At the highest dose of the RPH used, there was a significant increase in macrophage phagocytic percentage as well as the phagocytosis index, both of which were similar to levels found in normal mice. Since phagocytosis is an important action in initiating the innate immune response, the RPH may be an important agent in cancer therapy. The RPH also showed antioxidative effects judging by the increased activity of serum superoxide dismutase (SOD), which may contribute to the observed anticancer effects. Superoxide scavenging is believed to be the first line of defense against oxidative stress because it can be converted to other forms of toxic ROS. RPH also reduced plasma levels of malondialdehyde (a product of lipid peroxidation) in the tumor-bearing mice. Therefore, in addition to the enhanced immune system, the RPH seems to have exerted the anticancer effects through its radical scavenging (increased SOD levels) and antioxidative (decreased lipid peroxidation) properties.

Other peptides that have been reported to possess potential anticancer properties include:

- Small peptides (di- and tripeptides) obtained from bovine a-lactalbumin significantly increased proliferation of human peripheral blood lymphocytes and enhanced resistance to bacterial infection.
- Peptides in yogurt preparations were found to decrease cell proliferation, which may partially explain why consumption of yogurt has been associated with a reduced incidence of colon cancer. The mechanism of the antiproliferation effect is believed to involve binding of the peptides to opiate receptors on the cell.
- Some milk-derived peptides have also been shown to inhibit proliferation of leukemia cells, through increased apoptosis.
- A pentapeptide (EQRPR) obtained from enzymatic hydrolysis of rice bran was shown to be an effective inhibitor of cellular proliferation in colon, breast, liver, and lung cancer cell lines.

3.4.8   Antithrombotic Peptides

These are peptides that inhibit blood clotting and may be used to treat thrombosis (local coagulation of blood in the circulatory system). For example, bovine casein-derived peptide sequence called casoplatelin (11 amino acid residues) has been shown to inhibit both aggregation of ADP-activated platelets and binding of human fibrinogen to its receptor region on the platelets' surface. A smaller fragment (5 amino acid residues) called casopiastrin was obtained from digestion of casein with trypsin. Casopiastrin exhibited antithrombotic activity by inhibiting fibrinogen binding to thrombin during blood clotting. Antithrombotic peptides have also been derived from the milk protein of other species. For example, pepsin digests of sheep and human lactoferrin produced peptides that inhibited thrombin-induced platelet aggregation.

Bibliography

Alemán, A., E. Pérez-Santín, S. Bordenave-Juchereau, I. Arnaudin, M.C. Gómez-Guillén, and P. Montero. 2011. Squid gelatin hydrolysates with antihypertensive, anticancer and antioxidant activity. *Food Research International* 44: 1044–1051.

Aluko, R.E. 2008. Antihypertensive properties of plant-derived inhibitors of angiotensin I-converting enzyme activity: A review. *Recent Progress in Medicinal Plants* 22: 541–561.

Aluko, R.E. 2010. Food protein-derived peptides as calmodulin inhibitors. In *Functional food proteins and peptides*, IFT symposium series, ed. E.C.Y. Li-Chan, Y. Mine, and B. Jiang, 55–65. New York: Wiley-Blackwell.

Bougates, A., N. Nedjar-Arroume, L. Manni, R. Ravallec, A. barkia, D. Guillochon, and M. Nasri. 2010. Purification and identification of novel antioxidant peptides from enzymatic hydrolysates of sardinelle (*Sardinella aurita*) by-products proteins. *Food Chemistry* 118: 559–565.

Cao, W., C. Zhang, P. Hong, H. Ji, and J. Hao. 2010. Purification and identification of an ACE inhibitory peptide from the peptic hydrolysate of *Acetes chinensis* and its antihypertensive effects in spontaneously hypertensive rats. *International Journal of Food Science and Technology* 45: 959–965.

Chiu, L.-H., G.-S.W. Hsu, and Y.-F. Lu. 2006. Antihypertensive capacity of defatted soft-shelled turtle powder after hydrolysis by gastrointestinal enzymes. *Journal of Food Biochemistry* 30: 589–603.

De Moreno de LeBlanc, A., C. Matar, E. Farnworth, and G. Perdigon. 2006. Study of cytokines involved in the prevention of a murine experimental breast cancer by kefir. *Cytokines* 34: 1–8.

Girgih, A.T., C.C. Udenigwe, H. Li, A.P. Adebiyi, and R.E. Aluko. 2011. Kinetics of enzyme inhibition and antihypertensive effects of hemp seed (*Cannabis sativa* L.) protein hydrolysates. *Journal of the American Oil Chemists Association* 88: 1767–1774.

Huang, W., S. Chakrabarti, K. Majumder, Y. Jiang, S.T. Davidge, and J. Wu. 2010. Egg-derived peptide IRW inhibits TNF-α-induced inflammatory response and oxidative stress in endothelial cells. *Journal of Agricultural and Food Chemistry* 58: 10840–10846.

Huang, W.-H., J. Sun, H. He, H.-W. Dong, and J.-T. Li. 2011. Antihypertensive effect of corn peptides, produced by a continuous production in enzymatic membrane reactor, in spontaneously hypertensive rats. *Food Chemistry* 128: 968–973.

Ichimura, T., A. Yamanaka, T. Otsuka, E. Yamashita, and S. Maruyama. 2009. Antihypertensive effect of enzymatic hydrolysate of collagen and Gly-Pro in spontaneously hypertensive rats. *Bioscience, Biotechnology, and Biochemistry* 73: 2317–2319.

Ishiguro, K., Y. Sameshima, T. Kume, K. Ikeda, J. Matsumoto, and M. Yoshimoto. 2012. Hypotensive effect of a sweet potato protein digest in spontaneously hypertensive rats and purification of angiotensin I-converting enzyme inhibitory peptides. *Food Chemistry* 131: 774–779.

Je, J.-Y., K.-H. Lee, M.H. Lee, and C.-B. Ahn. 2009. Antioxidant and antihypertensive protein hydrolysates produced from tuna liver by enzymatic hydrolysis. *Food Research International* 42: 1266–1272.

Jung, E.Y., H.-S. Lee, H.J. Lee, J.-M. Kim, K.-W. Lee, and H.J. Suh. 2010. Feeding silk protein hydrolysates to C57BL/KsJ-db/db mice improved blood glucose and lipid profiles. *Nutrition Research* 30: 783–790.

Kannan, A., N.S. Hettiarachchy, J.O. Lay, and R. Liyanage. 2010. Human cancer cell proliferation inhibition by a pentapeptide isolated and characterized from rice bran. *Peptides* 31: 1629–1634.

Katayama, S., X. Xu, M.Z. Gan, and Y. Mine. 2006. Antioxidative stress activity of oligophosphopeptides derived from hen egg yolk phosvitin in Caco-2 cells. *Journal of Agricultural and Food Chemistry* 54: 773–778.

Kizawa, K., K. Naganuma, and U. Murakami. 1995. Calmodulin-binding peptides isolated from α-casein peptone. *The Journal of Dairy Research* 62: 587–92.

Ko, S.-C., J.-K. Lee, H.-G. Byun, S.-C. Lee, and Y.-J. Jeon. 2012. Purification and characterization of angiotensin I-converting enzyme inhibitory peptide from enzymatic hydrolysates of *Styela clava* flesh tissue. *Process Biochemistry* 17: 34–40.

Lee, H.-S., H.J. Lee, and H.J. Suh. 2011. Silk protein hydrolysate increases glucose uptake through up-regulation of GLUT 4 and reduces expression of leptin in 3T2-L1 fibroblasts. *Nutrition Research* 31: 937–943.

Li, H., N. Prairie, C.C. Udenigwe, A.P. Adebiyi, P. Tappia, H.M. Aukema, P.J.H. Jones, and R.E. Aluko. 2011. Blood pressure lowering effect of a pea protein hydrolysate in hypertensive rats and humans. *Journal of Agricultural and Food Chemistry* 59: 9854–9860.

Nakamura, Y., N. Yamamoto, K. Sakai, A. Okubo, S. Yamazaki, and T. Takano. 1995. Purification and characterization of angiotensin I-converting enzyme inhibitors from sour milk. *Journal of Dairy Science* 78: 777–783.

Ndiaye, F., T. Vuong, J. Duarte, R.E. Aluko, and C. Matar. 2012. Anti-oxidant, anti-inflammatory and immunomodulating properties of an enzymatic protein hydrolysate from yellow field pea seeds. *European Journal of Nutrition* 51: 29–37.

Omoni, A., and R.E. Aluko. 2006. Effect of cationic flaxseed protein hydrolysate fractions on the in vitro structure and activity of calmodulin-dependent endothelial nitric oxide synthase. *Molecular Nutrition & Food Research* 50: 958–966.

Onishi, K., N. Matoba, Y. Yamada, N. Doyama, N. Maryuyama, S. Utsumi, and M. Yoshikawa. 2004. Optimal designing of β-conglycinin to genetically incorporate RPLKPW, a potent anti-hypertensive peptide. *Peptides* 25: 37–43.

Pan, D., Y. Luo, and M. Tanokura. 2005. Antihypertensive peptides from skimmed milk hydrolysate digested by cell-free extract of *Lactobacillus helveticus* JCM1004. *Food Chemistry* 91: 123–129.

Saiga, A., S. Tanabe, and T. Nishimura. 2003. Antioxidant activity of peptides obtained from porcine myofibrillar proteins by protease treatment. *Journal of Agricultural and Food Chemistry* 51: 3661–3667.

Tsai, J.-S., T.-J. Chen, B.S. Pan, S.-D. Gong, and M.-Y. Chung. 2008. Antihypertensive effect of bioactive peptides produced by protease-facilitated lactic acid fermentation of milk. *Food Chemistry* 106: 552–558.

Udenigwe, C.C., and R.E. Aluko. 2011. Chemometric analysis of the amino acid requirements of antioxidant food protein hydrolysates. *International Journal of Molecular Sciences* 12: 3148–3161.

Udenigwe, C.C., and R.E. Aluko. 2012. Food protein-derived bioactive peptides: Production, processing and potential health benefits. *Journal of Food Science*. doi:10.1111/j.1750-3841.2011.02455.x.

Xue, Z., W. Yu, M. Wu, and J. Wang. 2009. In vivo and antioxidative effects of a rapeseed meal protein hydrolysate on an S180 tumor-bearing murine model. *Bioscience, Biotechnology, and Biochemistry* 73: 2412–2415.

Zhang, J.-H., E. Tatsumi, C.-H. Ding, and L.-T. Li. 2006. Angiotensin I-converting enzyme inhibitory peptides in douchi, a Chinese traditional fermented soybean product. *Food Chemistry* 98: 551–557.

Bioactive Polyphenols and Carotenoids

4.1 Introduction

Polyphenolic compounds are commonly found in virtually all types of food plants and constitute an important component of the human diet. Traditionally, the polyphenols were viewed from the angle of their adverse effects on human health because of their ability to bind and insolubilize various nutrients such as minerals (especially heme iron), proteins, and carbohydrates. Though ascorbic acid can counteract iron inhibition by low levels of polyphenols, it is ineffective in the presence of high levels of polyphenols. Therefore, high levels of polyphenolic compounds in foods are sometimes believed to be responsible for decreased food digestibility and reduced nutrient bioavailability. However, recent research trends have identified certain food polyphenols as being potential health-promoting agents because of their ability to act as antioxidants and free radical scavengers. For example, as anticancer agents, polyphenols act by removing carcinogenic agents (through chelation), modulating cancer cell signaling and cell cycle progression, inducing various enzyme activities, and promoting apoptosis. Polyphenols can enhance antioxidant enzyme levels such as NADPH-quinone oxidoreductase, glutathione S-transferase, cytochrome P450, glutathione peroxidase, and catalase, which increases capacity of the body to detoxify carcinogens. Polyphenols downregulate the p38/CREB signaling and inhibit protein phosphorylation (blocks cell cycle progression) and COX-2 expression to attenuate cancer cell growth. In relation to cardiovascular diseases, polyphenols can improve blood circulation by decreasing plasma cholesterol and free fatty acids. For example, during myocardial infarction, pretreatment with caffeic acid resulted in decreased activity of the 3-hydroxy-3-methylglutaryl-coenzyme A reductase, the enzyme that catalyzes the rate-determining step during hepatic cholesterol biosynthesis. Other beneficial effects of polyphenols can be illustrated by the following demonstrated activities.

1. Polyphenol-rich blackberry extract exhibited protective effects against oxidative stress in carbon tetrachloride-treated rats by reducing lipid peroxidation in the liver through increased activities of antioxidant enzymes like catalase, glutathione peroxidase, and superoxide dismutase.

2. Polyphenolic extracts from berries (blueberry, blackberry, strawberry, raspberry, and cranberry) and grapes are effective scavengers of carbonyl compounds that have been implicated in the production of advance glycation end products (AGEs). In chronic disease conditions such as diabetes, aging, and some cancers, reactive carbonyl groups such as methylglyoxal and glyoxal react with proteins in a similar way as the in vitro Maillard reaction to form glycated proteins. The AGEs are polymeric proteins that have been cross-linked through intra- and intermolecular bonds, which render the protein nonfunctional. The most common AGE compound in the body is glycated hemoglobin (HbA1c) whose level in the blood can

R.E. Aluko, *Functional Foods and Nutraceuticals*, Food Science Text Series, DOI 10.1007/978-1-4614-3480-1_4, © Springer Science+Business Media, LLC 2012

be used as an indication of long-term glucose control as well as the risks for complications in diabetic patients. It is now known from in vitro work that berry procyanidins, especially catechin, can react with methylglyoxal to form inactive adducts, which seems to be the main mechanism by which berry extracts prevent protein glycation.

3. Anthocyanin-rich mulberry extract effectively prevented atherosclerosis, inhibited gastric cancer cell through increased apoptosis, and inhibited metastasis of melanoma cells. The mechanism of anticancer effects was due to increased activation of caspase-3, which induces apoptotic cell death. Likewise, sweet potato anthocyanins also reduced hepatic lipid accumulation and weight gain through upregulation of adenosine monophosphate-activated protein kinase (AMPK). AMPK downregulates gene expression of lipid synthesis agents such as sterol regulatory element-binding protein 1 and its target genes such as acetyl-coenzyme A carboxylase and fatty acid synthase. The net effect is increased fatty acid oxidation, reduced lipid storage, and reduced risk of developing atherosclerosis.

4. The leaf of oil palm is rich in catechins whose extract has been shown to have vasodilatory properties and could be an important agent for reducing blood pressure. In addition, the catechin-rich extract performed as an antioxidant by reducing LDL oxidation when tested in a cell culture model. The effects of the catechin-rich extract were reflected as attenuated organ injury, which led to improved performance of various organs. Under nitric oxide (NO) deficiency conditions, microvascular perfusion failure, portal hypertension, and increased hepatic resistance contribute to hepatic injury. The catechin-rich extract reduced hepatic injury that is associated with NO deficiency through its blood pressure-lowering effect as a result of enhanced NO production, which improves microvascular perfusion and reduces hepatic resistance. Kidney damage resulting from NO deficiency was also ameliorated by the catechin-rich extract as evidenced by the reduced level of blood urea nitrogen in oil palm catechin extract-fed rats, suggesting improved glomerular function. The catechin-rich extract also reduced renal level of malondialdehyde, a marker of lipid peroxidation, and suggests an important role for the antioxidant properties of catechins in reducing kidney damage.

5. Addition of blackberry extract to aging (19 months old) rat diet led to significant increase in memory scores, which suggests that age-related decline in object memory may be prevented or even reversed by blackberry treatment. The antioxidant ability of the blackberry polyphenols to reduce oxidative stress may be responsible for the observed effects.

Most plant polyphenols occur as simple phenols and flavonoids, which are compounds of relatively low molecular weights and are soluble (extractable) depending on their chemical structure and presence of polar hydroxyl and glycosyl groups. Other plant polyphenols are mostly high molecular polymers such as condensed tannins and proanthocyanidins, which are categorized as non-extractable polyphenols. Though not readily absorbed, the non-extractable polyphenols are believed to be 15–30 times more potent than soluble polyphenols at quenching peroxyl radicals. Thus, the non-extractable polyphenol polymers can act as antioxidants within the digestive tract to protect lipids, proteins, and carbohydrates from oxidative damage during digestion. In this way, the soluble polyphenols are spared for absorption into the blood circulatory system and transported to organs and tissues where they can exert physiological benefits as antioxidants, especially to protect vital cellular structures like the cell membrane bilayer and genetic materials. Absorption of these polyphenols has been confirmed through detection of their metabolites (usually conjugated to sulfate or glucuronic acid groups) in urine of human subjects that consumed polyphenolic-rich drinks. Figure 4.1 shows a structural representation of some common polyphenolic compounds.

In general, several soluble or extractable polyphenols can be metabolized within the gastrointestinal tract, though most of the glycosides require hydrolysis to produce the aglycones

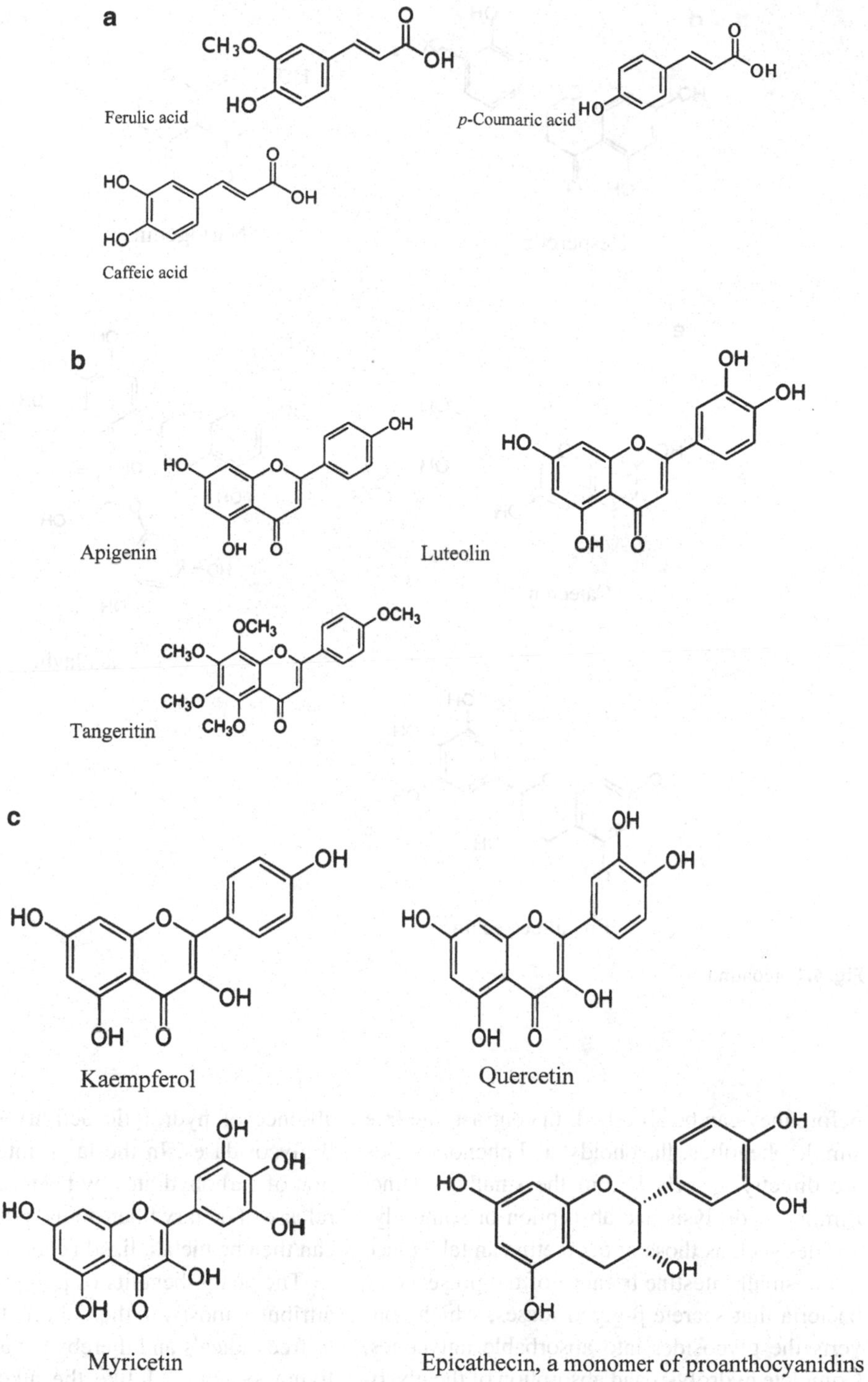

Fig. 4.1 (a) Simple phenolic acids. (b) Flavones. (c) Flavonols. (d) Flavanones. (e) Flavanols

d

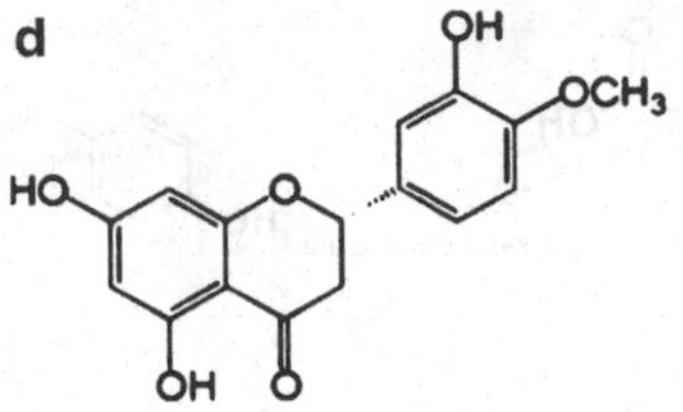

Hesperetin

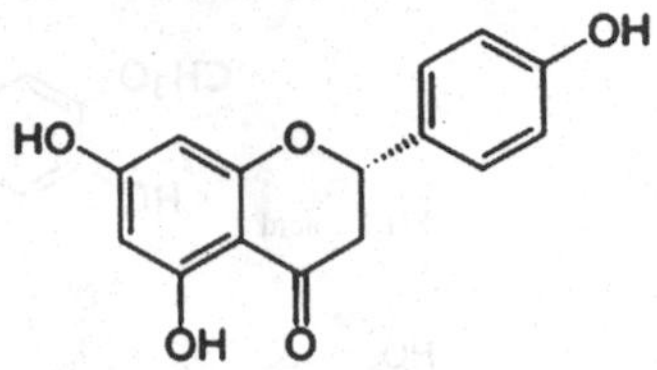

Naringenin

e

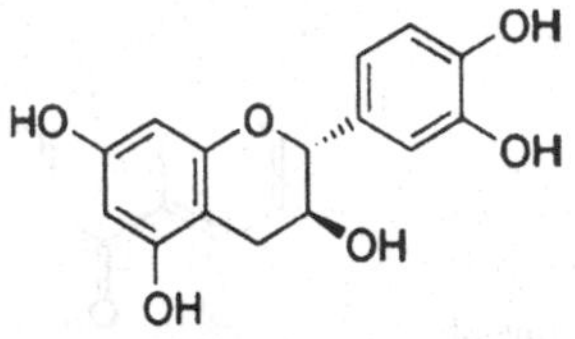

Catechin

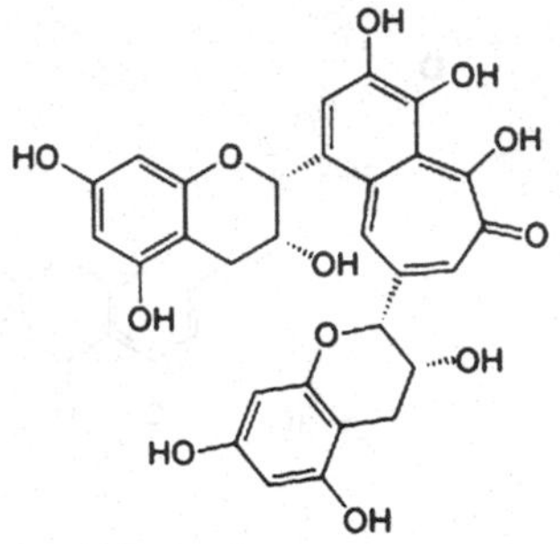

Theaflavin

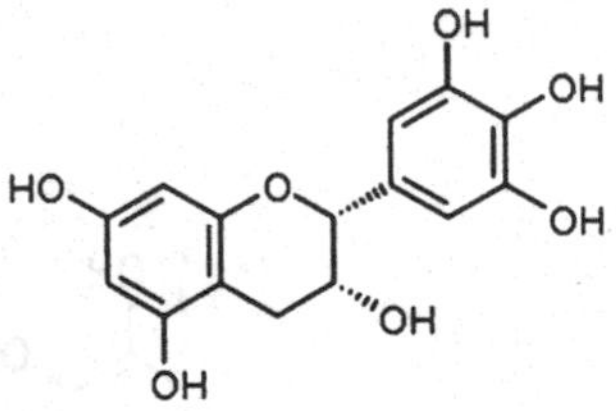

Epigallocatechin

Fig. 4.1 (continued)

before they can be absorbed. In contrast, the free simple phenolics, flavonoids, and phenolic acids are directly absorbed from the small intestine. Limited hydrolysis and absorption of some glycosides such as those of quercetin can take place in the small intestine because of the presence of bacteria that secrete β-glycosidases, which converts the glycosides into absorbable aglycones. Complete hydrolysis and absorption of the glycosides occur in the large intestine where bacteria β-glycosidases are more plentiful than in the small intestine. Confirmation of this process was obtained when germ-free rats were shown to excrete intact flavonoid glycosides due to the absence of hydrolytic activity of germ-derived β-glycosidases. In the large intestine, fermentation of carbohydrates by bacteria can lead to the release of dietary fiber-bound polyphenols, which can then be metabolized accordingly.

The health benefits of polyphenols have been attributed mostly to their ability to donate protons to free radicals and thereby act as antioxidants in living systems. Unlike the alkoxy radicals, the phenoxy radicals are relatively stable (Eqs. 4.1 and 4.2), which prevents initiation of a new chain of reaction. By reacting with other free radicals, the phenoxy radicals can also act as terminators of the propagation reactions (Eqs. 4.3 and 4.4).

However, it should be noted that under conditions of high concentrations of phenolics, high pH, and presence of iron, the phenolics can act as prooxidants by initiating free radical formation.

$$ROO^\bullet + PPH \rightarrow ROOH + PP^\bullet \quad (4.1)$$

$$RO^\bullet + PPH \rightarrow ROH + PP^\bullet \quad (4.2)$$

$$ROO^\bullet + PP^\bullet \rightarrow ROOPP \quad (4.3)$$

$$RO^\bullet + PP^\bullet \rightarrow ROPP \quad (4.4)$$

There has been a huge increase in food phenolics research because of their perceived ability to prevent chronic diseases such as cancer, kidney failure, hypertension, and other forms of cardiovascular diseases.

4.2 Structure-Function Considerations

The chemical structure of polyphenols dictates their antioxidant efficiency because phenol as a compound has no antioxidant property. Diphenols in the *ortho-* and *para-*forms possess antioxidant activity, which is enhanced when hydrogen atoms on the rings are substituted by ethyl or n-butyl groups. As an example of how structure dictates function, the presence of one or more of the following structural characteristics is known to enhance the antioxidant properties of flavonoids:
1. An *o*-diphenolic group in the B ring
2. A 4-oxo function that exists in conjugation with a 2,3 double bond
3. OH-groups in the 3rd and 5th positions

A flavonol such as quercetin is one of the most potent natural antioxidants because it contains all the above-named three structural characteristics. There is a direct correlation between the number of OH-groups in flavonoids and their antioxidant efficiency, but the presence of attached sugar molecules has a negative effect. Therefore, in general, glycosides have little or no antioxidant activity whereas upon hydrolysis by β-glycosidases, the resultant aglycones possess antioxidant properties. For example, consumption of prebiotics has been shown to increase the glycolytic capacity of colon microflora, which leads to enhanced conversion of glycosidic polyphenolic compounds to the aglycone forms. The increased microflora activity also leads to higher rate of metabolism of complex polyphenols into simpler and more readily bioavailable breakdown (low molecular weight) products. Therefore, dietary intervention that includes simultaneous administration of prebiotics and polyphenols could lead to greater biological effects of polyphenolic compounds.

4.3 Specific Polyphenolic Products

4.3.1 Grape and Red Wine Polyphenol Extracts

As antioxidants, red wine polyphenolic extract (RWPE) is able to reduce the level of superoxide anion in animal tissues and could be a useful therapeutic tool for the treatment and prevention of endothelial dysfunction. Presence of high levels of tissue reactive oxygen species (ROS) such as the superoxide anion is a known potentiator of cardiac hypertrophy. Consumption of RWPE led to significant reductions in cardiac hypertrophy that is associated with fructose-fed rats. RWPE also has vasorelaxing properties, possibly through enhanced production of nitric oxide which may be responsible for its antihypertensive properties. This is because red wine polyphenols can increase the nitric oxide-cyclic GMP (guanosine-3′,5′-monophosphate) pathway in vascular tissues. Nitric oxide is a well-known vasodilatory agent that contributes to maintenance of normal blood pressure. Red wine polyphenols can bind omega-3 better than omega-6 fatty acids, thus protecting the former from the damaging effects of ROS. By protecting omega-3 fatty acids from oxidation, the red wine polyphenols could help to increase level of anti-inflammatory eicosanoids in the plasma, reduce LDL oxidation, and lower the risk for development of atherosclerosis plaques.

The presence of different polyphenolics in food products may provide increased health benefits based on synergistic effects, when compared to the effects of single polyphenol compounds. For example, the antioxidant power of a fruit juice mixture that contains different proportions of juices from grape, cherry, blackberry, blackcurrant, and raspberry is almost 400% and 700% more than that of anthocyanins and catechins, respectively. The fruit juice mixture is also able to protect cells against peroxide-mediated lipid oxidation and cell damage. This is very important since lipid peroxides can damage cell membrane integrity resulting in altered fluidity and disruption of membrane structure and function. The lipophilic character of some of the phenolics may improve their ability to enter the cell or to become localized in lipid compartments in order to exert their beneficial effects. As antioxidants, the fruit juice polyphenols prevent rapid cellular utilization of glutathione, which is one of the important endogenous antioxidant molecules. By sparing glutathione, the polyphenols contribute to maintaining long-term antioxidant status and help in stabilizing cells against the damaging effects of endogenous free radicals and peroxides.

Red grape skin polyphenolic extract that is enriched in anthocyanins prevented development of hypertension and cardiac hypertrophy and formation of ROS when present as part of high-fructose diet. Similarly, a grape seed extract that is enriched in galloylated procyanidins prevented insulin resistance, hypertriglyceridemia, and overproduction of ROS. The grape skin and grape seed extracts both modulated expression of NADPH oxidase genes, which accounts for their ability to reduce production of ROS. Bioavailability of grape polyphenols has been demonstrated, and various metabolites (usually glucuronidated or sulfated) have been detected in urine following consumption of grape products.

Red grape juice is known to contain high levels of procyanidin B2 (PB2), a flavonoid that is also present in cocoa and red wine. It has been shown that PB2 (Fig. 4.2) may have a protective role in the prevention of oxidative stress-related intestinal injury and gut pathologies. Under regular hemostasis, the gastrointestinal tract (GIT) is constantly exposed to ROS that can cause severe damages and lead to pathogenesis of various chronic diseases, including cancer. Flavonoids

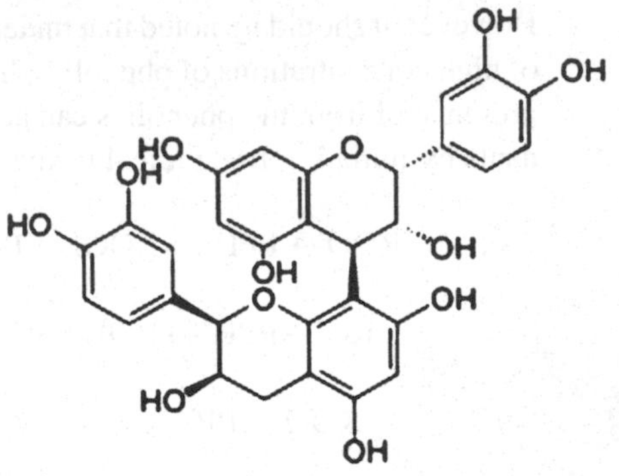

Fig. 4.2 Chemical structure of proanthocyanidin B2

are well known for their antioxidant and free radical scavenging activities, which can be exploited in protecting the GIT, especially the colon from the damaging effects of oxidative stress. For example, treatment of Caco-2 cells with PB2 led to increased expression levels of the antioxidant enzyme glutathione S-transferase P1 (GSTP1) in addition to increased translocation of nuclear factor erythroid 2-related factor 2 (Nrf2). Glutathione S-transferases (GSTs) are a group of enzymes that provide protection against free radicals and carcinogens by promoting enhanced excretion of xenobiotics, which are conjugated with glutathione to increase urinary solubility. The p class of GSTs has also been implicated as regulators of cell transformation and carcinogenesis. Nrf2 is a transcription activator that binds to antioxidant response (ARE) elements in the promoter regions of target genes and is an important factor for the coordinated upregulation of genes in response to oxidative stress. Activity (nuclear translocation) of Nrf2 requires protein phosphorylation by cellular kinases, including mitogen-activated protein kinases (MAPKs) and phosphatidylinositol 3-kinase (PI3K). Therefore, compounds that activate cellular kinases will enhance Nrf2 activity, which leads to improved cellular antioxidant and detoxification capacities. Pretreatment of Caco-2 cells with PB2 prior to induction of oxidative stress led to significant increases in level of GSTP1, which was accompanied by attenuation of oxidative damage and cell death. PB2 treatment also increased the

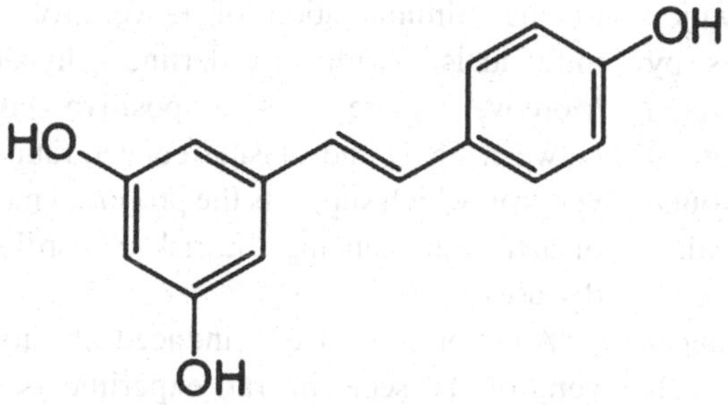

Fig. 4.3 Structure of resveratrol

nuclear protein levels of Nrf2 for up to 20 h as well as the phosphorylated protein levels of extracellular signal-regulated protein kinases (ERKs) and p38 MAPK. Caco-2 cells treated with PB2 and peroxide showed absence of oxidative radicals and suppression of peroxide-induced cell death. In contrast, cells treated with peroxide alone showed high levels of free radicals and reduced cell viability. When specific inhibitors of ERKs and p38 were added to the PB2 treated cells, there was no protective effect because high levels of free radicals and reduced cell viability were observed. Thus, it is evident that PB2 works as an antioxidant and protects cells against oxidative injury through induction of the ERKs and p38 signaling pathways.

4.3.2 Resveratrol (3,5,4′-Trihydroxystilbene)

Resveratrol (Fig. 4.3) is a non-flavonoid polyphenolic compound found in the skin of dark-colored grapes and products made from such grapes, such as wine and grape juice.

It belongs to the stilbene class of aromatic phytochemicals and occurs as a free (*cis-* or *trans*-configurations) aglycone (less soluble but the active form) or as a glycoside (called piceid, the more soluble transport form). Resveratrol is transferred across the intestinal border and into the blood circulatory system as a glucuronide (glycoside) but may be cleaved back into the aglycone form once it reaches the organs or body fluids where it is acted upon by β-glucuronidases. The compound is a known inhibitor of cyclooxy-genase-1 and cyclooxygenase-2 (COX-1 and COX-2), which are usually overexpressed in colon cancer. Resveratrol also inhibits monoamine oxidase while promoting protease degradation of Aβ and acting as an efficient antioxidant. In mice experiments, oral administration of resveratrol induced both angiogenesis and neurogenesis in the hippocampus through stimulation of sensory neurons in the gastrointestinal tract. Spatial learning in the Morris water maze was also improved. The effects of resveratrol on memory are mediated through increased release of calcitonin gene-related peptide (CGRP) and insulin-like growth factor-1 (IGF-1) in the mice hippocampus. CGRP functions as a transmitter in the pathway of the sensory nervous relay system while IGF-1 enhances synaptic transmission and plasticity. These activities are part of the reasons for the use of resveratrol as an agent against Alzheimer's disease (AD) and for improvement of cognitive functions.

Resveratrol has other biological activities such as inhibition of cell proliferation and induction of apoptosis in various cancer cells; in animals and humans, resveratrol acts to prevent colon cancer. The anti-colon cancer effect of resveratrol may be attributed to its ability to inhibit a specific signaling pathway, the Wnt pathway that is activated in most colon cancers. Resveratrol has also been shown to inhibit the occurrence and progression of other types of cancers such as mammary and skin tumors. There are contradictory reports that suggest lack of resveratrol activity against these types of cancers. In cases where anticancer effects were demonstrated, the activity is through several mechanisms that involve anti-initiation and induction of terminal differentiation of cells that prevents further cell division. The anti-initiation effect is due primarily to the antioxidant and anti-mutagenic effects, inhibition of bioactivation of various carcinogens as well as induction of enzyme(s) (e.g., quinine reductase) involved in phase II drug metabolism. The dimer form of resveratrol is called viniferin, and both forms have strong inhibitory effects towards various cytochromes such as p1A1, 1A2, 1B1, 2A6, 2B6, 2E1, 3A4, and 4A that are involved in bioactivation or deactivation of various carcinogens. Resveratrol has shown antiangiogenesis activities, which could contribute to the anticancer effects by preventing vascularization of tumors. However, a major drawback is that resveratrol inhibits normal physiological angiogenesis (e.g., during wound healing) as well as pathological angiogenesis during tumor growth.

Other mechanisms involved in the anticancer effects of resveratrol include reduction in the cellular levels of ornithine decarboxylase, an enzyme that is important for the synthesis of bioamines and cell proliferation. Resveratrol can regulate the expression of several apoptotic factors such as p21, p53, and *Bax*. The p21 factor is involved in the arrest of cell cycle progression in newly formed cancerous cells. The presence of resveratrol can increase the expression of p53, a tumor suppressor gene, and also lead to increases in caspase-mediated and CD95 signaling-dependent apoptosis. Also significant is that resveratrol has been found to inhibit enzymes such as DNA polymerase and ribonucleotide reductase that provide proliferating cells with deoxyribonucleotides needed for synthesis of DNA molecules. Lack of DNA molecules will ultimately lead to loss in ability of the cells to divide since the *S phase* cannot be completed. Resveratrol has estrogenic activity as a result of structural similarity with diethylstilbestrol and could behave as competitive inhibitor of natural estrogens. For example, *trans*-resveratrol prevents binding of estradiol to type I estrogen receptors in human breast cancer cells and in this way can prevent development of certain types of breast cancer that is caused by high levels of estradiol or similar hormones. In men, resveratrol may be able to prevent prostate cancer formation because the compound inhibits androgen-stimulated cell growth and androgen upregulated genes.

Resveratrol has also been implicated in the cardiovascular health benefits such as increased blood flow, decreased inflammation, and decreased oxidative stress that are associated with consumption of grape products. This is because resveratrol upregulates eNOS, which leads to increases in NO-mediated vasodilation and blood flow. Endothelial function measured by flow-mediated vasodilation (FMD) was increased following administration of resveratrol to obese individuals with borderline hypertension. Moreover, there was a positive correlation between FMD and plasma concentration of resveratrol, which supports the proposed mechanism of action in reducing the risk of cardiovascular diseases.

Amelioration of diet-induced obesity by resveratrol as seen in rat experiments may be exploited as a means of improving glucose metabolism and as a therapeutic aid for diabetic patients. However, there is lack of information on the ability of resveratrol to induce weight loss in human subjects. Few human trials have demonstrated blood-reducing effects of resveratrol, but more data is required before definitive conclusions can be made. The ability of resveratrol to improve glucose metabolism is believed to be as a result of various mechanisms such as anti-inflammatory action and inhibition of protein-tyrosine phosphatase 1B, the enzyme that inactivates insulin receptors. Administration of 5 mg resveratrol twice daily to type 2 diabetic men over a 4-week period led to improved insulin sensitivity, reduced blood glucose, and delayed glucose peak following consumption of a standardized meal. Resveratrol is attractive as a safe antidiabetic product without negative side effects because even at high doses, no incidence of hypoglycemia has been reported. The beneficial effect on weight is partly due to the ability of resveratrol to inhibit fatty acid and triglyceride syntheses as well as increase phosphorylation and activation of AMPK. AMPK acts by upregulating fatty acid oxidation and increasing Glut 4 translocation-dependent glucose uptake. Glucose uptake is also increased through upregulation of estrogen receptor-α, which in turn increases the expression of Glut 4 through the phosphatidylinositol 3-kinase (PI3K) and AKT pathway. Reductions in fat synthesis seem to be mediated through downregulation of lipogenesis genes. Resveratrol also increases mitochondrial biogenesis and oxidative phosphorylation and contributes to suppression of lipid accumulation in the adipose tissue through upregulation of SIRT 1 (an enzyme that inhibits adipocyte differentiation and triglyceride accumulation) and peroxisome proliferator-activated receptor gamma (PPARγ) coactivator (PGC)-1α. This is because PGC-1α is a mediator for increased mitochondrial biogenesis and oxidative metabolism, which enhances fatty acid oxidation to produce decreased adipose tissue weight. Therefore, resveratrol can be used as an agent for the prevention of obesity and diabetes as evident from animal feeding trials that showed effective reductions in fat depot sizes, visceral fat, liver weight, and total body fat. While moderate uses of resveratrol have been shown to be safe, oral administration of large doses (3 g/kg body weight per day) or 14 mg/L in drinking water have been shown to cause oxidative damages to the kidneys and liver.

4.3.3 Apple Polyphenols

While there has been no definite data from human studies on the specific health benefits of apple polyphenols, several animal and cell culture experiments suggest consumption of apples could reduce the risk of several chronic diseases. The health benefits of apples are related mainly to the high content of polyphenolic compounds (especially flavonoids), which can vary from ~660 to 2120 mg/kg fresh fruit weight, values that are much higher than the combined contents of vitamin C, magnesium, and calcium. Higher levels of the polyphenols including some anthocyanins are present in the apple skin when compared to the fruit tissue. It is well known in general that fruit flavonoids can act as antioxidants and in the process spare vitamin C, reduce inflammation, reduce cell proliferation, and inhibit blood clot formation. Apple quercetin and procyanidins have been shown to reduce levels of some carcinogenesis markers and enhance the vascular system, respectively. Quercetin has protective effects against H_2O_2-induced neurodegradation and may offer protective effects against stress-induced neurotoxicity such as Alzheimer's disease. Treatment of liver cancer cells with the polyphenol-rich apple peel was effective in reducing cell growth and proliferation. Apple extracts prevented mammary cancer development in experimental rats at doses

equivalent to human consumption of 1, 3, or 6 apples per day with tumor incidence reduced by 17%, 39%, and 42%, respectively. Within the gastrointestinal tract, apple polyphenols exert direct effects such as iron sequestration and scavenging of reactive nitrogen, oxygen, and chlorine species. The antioxidant health benefit of apple consumption was demonstrated in obese rats with data showing that heart and urine concentrations of malondialdehyde (product of lipid peroxidation) were reduced. In rat experiments, apple polyphenolic extract also reduced lipid peroxidation that is associated with high dietary cholesterol. However, it has also been suggested that the increase in human plasma antioxidant activity may not be due to apple polyphenols alone but as a result of fructose-induced formation of urate, an important physiological antioxidant. The higher contents of available polyphenols in apple products that contained the skin when compared to apple juice suggests that polyphenols play an important role in the antioxidant and anti-inflammatory functions of apple fruits. Therefore, it has been recommended that whole apples that include the skin be consumed to obtain maximum health benefits from the bioactive properties of the polyphenols.

4.3.4 Lychee Fruit Polyphenols

The lychee tree is a tropical and subtropical fruit plant that is native to Southern China and Southeast Asia but is now cultivated in several parts of the world. The fleshy fruit has a pink-red outer skin that is rich in polyphenols. Through proprietary methods, a polyphenolic-rich product called Oligonol has been produced from lychee fruits. Oligonol contains a mixture of polyphenolic compounds of varying sizes: 15.7% as monomers mostly (+)-catechin and (-)-epicatechin; 13% as dimers (proanthocyanidins); and 71.3% as other polyphenolics ranging from trimers to polymers. Polyphenolic compounds present in Oligonol have been shown to be highly bioavailable, which contribute to enhanced physiological properties of the product. During in vitro tests, Oligonol inhibited xanthine oxidase activity in a dose-dependent

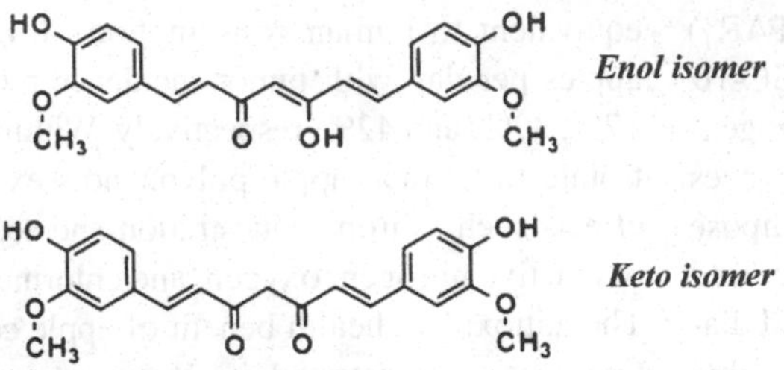

Fig. 4.4 Structure of two main isomers of curcumin

manner. In human experiments, consumption of Oligonol led to significant reductions in serum and urine uric acid contents, an indication of reduced synthesis of uric acid arising from inhibition of xanthine oxidase activity. Thus, the Oligonol polyphenols may be suitable for use in reducing the risk of hyperuricemia and gout.

4.3.5 Curcumin

The rhizome of the perennial herb Curcuma longa is usually ground into spice turmeric, which contains curcumin as the major bioactive polyphenol. Curcumin (Fig. 4.4) has several biological activities such as chemopreventive, anticancer, antiangiogenesis, anti-inflammatory, antioxidant, and chemotherapeutic. However, while there has been human intervention trials that looked at the anti-inflammatory and anticancer activities, most of the studies on obesity have been performed in animal (mice and rats) experiments.

Addition of curcumin to a rat diet led to increased bile acid excretion and reductions in liver cholesterol, liver weight, plasma free fatty acids, plasma triglycerides, and total body weight. An important biological attribute of curcumin is the ability to reduce angiogenesis (formation of new blood vessels), which is vital for growth and expansion of adipose tissues and cancerous tumors. Mechanism of action involves downregulation of important cellular factors such as vascular endothelial growth factor (VEGF), basic fibroblast growth factor (bFGF), epidermal growth factor (EGF), angiopoietin, and hypoxiainducible factors (HIF)-1α. Curcumin enhanced the level of activated AMPK while reducing acetyl-coenzyme A carboxylase (ACC) activity,

which leads to carnitine palmitoyl transferase 1 upregulation and increased mitochondria oxidation of long-chain fatty acids. In rats fed curcumin-supplemented high-fat diet, there was reduced intensity of fatty liver formation and reduced fatty acid and cholesterol synthesis in addition to increased hepatic β-oxidation. Curcumin can also attenuate disease progression in type 2 diabetes and liver inflammation as evidenced by the lower expression levels of nuclear factor kappa B (NF-κB) and reduced migration of macrophages into the adipose tissue. Curcumin inhibits expression of PPARγ and C/EBPα, the key transcription factors involved in adipogenesis and lipogenesis in the adipose tissue. By suppressing differentiation of preadipocytes to adipocytes, curcumin attenuated growth and expansion of adipose tissue.

4.3.6 Phytosterols

These are compounds that have similar chemical structure to cholesterol but are found only in plant-based foods such as vegetables, fruits, seeds, and vegetable oils. There are two main forms of phytosterols: sterols (Fig. 4.5a, b) have unsaturated aromatic ring structure while stanols (Fig. 4.5c, d) are the saturated forms. The mechanism of action of phytosterols is believed to be through competitive exclusion of cholesterol from bile acid micelles and/or through an induction of cholesterol precipitation in the intestinal tract, which limits cholesterol absorption.

Due to their lipophilic properties, phytosterols are more physiologically effective (i.e., reduction in blood cholesterol levels) when they are consumed in the form of an edible fat/oil. Fortification of margarine with phytosterols has been one of the most popular ways of developing cholesterollowering functional foods. In fact, human trials have shown that a combination of cholesterollowering drugs and cholesterol-lowering margarine can provide an additional 10–20% reduction in low-density lipoprotein cholesterol (LDL-C). Therefore, the combinations of drugs and functional food phytosterols may be used as a form of treatment for hypercholesterolemia. Apart from margarines (and other types of spreads),

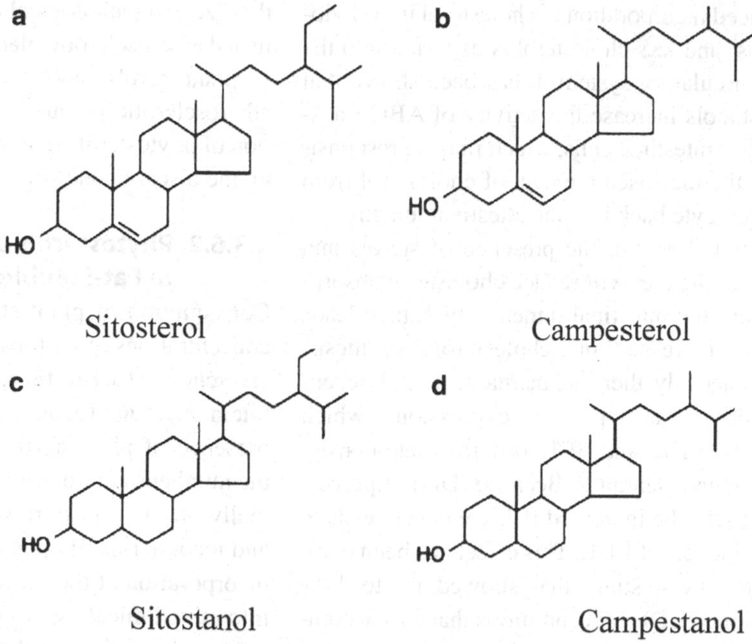

Fig. 4.5 Structures of typical phytosterols (**a, b**) and phytostanols (**c, d**)

phytosterol-fortified beverages have also been shown to lower serum cholesterol by as much as 8% in human subjects. Hypercholesterolemic patients that consumed phytosterol-fortified milk have also been found to show as much as 12% reduction in blood cholesterol; healthy individuals showed similar results with up to 11% lower content of serum cholesterol. Thus, it is evident that the cholesterol-lowering benefit of phytosterols is similar in normo- and hypercholesterolemic people. However, it should be noted that dietary intake of phytosterol given as small doses throughout the day is more efficient at lowering blood cholesterol than when given as one single large dose. Phytosterols are also relatively safe even at high doses, as evidenced by the fact that even though a phytosterol diet affected cholesterol hemostasis in mice brain, there was no effect on memory or anxiety-related behavior.

4.3.6.1 Phytosterols and Intestinal Absorption of Cholesterol

Intestinal absorption of plant sterols is about 0.4–3.5%, which is higher than that of stanols (0.02–0.3%) but much lower than that of choles-

terol (35–70%). In general, serum concentrations of plant sterols are 10–30 times higher than those of plant stanols, which is a reflection of the differences in their rates of absorption from the digestive tract. Interactions between sterols and stanols can lead to reduced absorption of each when present together in foods. It is believed that the poor absorption properties of plant sterols and stanols are due to their poor esterification, which limits incorporation into chylomicrons. Chylomicrons provide the vehicle by which cholesterol and the phytosterols are transported into blood circulation and subsequently into the liver. Several mechanisms have been proposed to explain the cardioprotective properties of dietary sterols and stanols.

Absorption of cholesterol is enhanced when present in mixed micelles with mono- and diglycerides, fatty acids, phospholipids, and bile salts. Displacement of cholesterol from mixed micelles by the more hydrophobic sterols and stanols reduces micellar cholesterol concentrations, which lowers absorption rates. It is also believed that sterols and stanols may reduce the rate of cholesterol esterification in the enterocyte, which then leads to reduced incorporation of cholesterol into chylomicrons, and less cholesterol is exported into the blood circulatory system. It has been shown that plant stanols increase the activity of ABC transporters in intestinal cells, which may be responsible for the increased removal of cholesterol from the enterocyte back into the intestinal lumen.

As noted above, the presence of sterols and stanols in the diet will reduce cholesterol absorption from the intestinal lumen, which then leads to an increase in cholesterol synthesis. Simultaneously, there are increases in LDL receptor mRNA and protein expression, which increases LDL and IDL (intermediate-density lipoproteins) clearance. Because IDL is a precursor of LDL, the increased IDL clearance leads to reduced levels of LDL. This effect has been demonstrated by a study that showed up to 14% reduction in LDL concentrations that was accompanied by lower cholesterol absorption, higher LDL receptor expression, and higher endogenous cholesterol synthesis following a daily intake of 2.0–2.5 g of plant sterols and stanols. Ingestion of plant sterols and stanols had no effect on plasma triacylglycerol and HDL (high-density lipoproteins) cholesterol levels.

Normally, rodents have plasma lipoprotein profiles that are different from that of human beings and are unsuitable for mimicking human atherosclerotic lesion development. Moreover, rodents are very resistant to the development of atherosclerosis; therefore, transgenic rodents are the most suitable alternatives for evaluating the cardioprotective effects of plant sterols and stanols. In ApoE-deficient mice, atherosclerotic plaques can be produced upon feeding of high-cholesterol diets and can be used to evaluate the effect of phytosterols on atherosclerosis. Addition of mixtures of sterols and stanols to diets can lead to reductions in atherosclerotic lesion size and complexity in ApoE-deficient mice, effects that are due to the significant reductions in plasma cholesterol levels. Consumption of plant sterol is known to prevent plaque formation in ApoE-deficient mice through reductions in the concentration of atherogenic β-VLDL (very-low-density lipoproteins) particles, which leads to decreased formation of foam cells. However, it seems as if the size and pathological characteristics of established atherosclerotic plaques cannot be modified by plant sterols and stanols. Therefore, once the atherosclerotic plaques have formed, consumption of phytosterols may have no beneficial effect on the disease condition.

4.3.6.2 Phytosterols and Absorption of Fat-Soluble Antioxidants

Consumption of plant stanols can lower serum concentrations of various lipophilic hydrocarbon carotenoids (α-carotene, β-carotene, lycopene, lutein, etc.) and tocopherols. This is because the presence of phytosterols leads to reductions in the number of circulating LDL particles that normally act as transport vehicles for carotenoids and tocopherols. Phytosterols also interfere with incorporation of the carotenoids and tocopherols into mixed micelles, which reduces the possibility for absorption of these lipophilic hydrocarbon carotenoids. Therefore, a potential side effect of phytosterol consumption is decrease in the antioxidant capacity of blood and tissues. Fortification with the lipophilic hydrocarbon carotenoids may be a suitable means of avoiding low-serum antioxidant capacity during regular consumption of phytosterol-containing foods. Effects of phytosterols on the absorption of vitamins A and D seem negligible probably because the two vitamins are not dependent on LDL for transportation. In contrast, phytosterol effect on vitamin K is not clear because of contradictory results from various experiments. However, short-term experiments have shown that daily consumption of plant stanols had no effect on warfarin anticoagulant therapy, suggesting that vitamin K levels were not significantly changed by the stanols.

4.3.6.3 Phytosterols and Membrane Integrity

Regular consumption of phytosterols may have deleterious effects on the integrity of cell membranes. This is because dietary phytosterols become incorporated into cell membranes where they can change membrane properties. In stroke-prone spontaneously hypertensive rats, consumption of phytosterols caused early mortality as a result of increased fragility and reduced deformation ability of red blood cells, which severely limited their ability to pass through the smallest capillaries. Consumption of phytosterols caused increased stiffness in rat red cell membranes, which can lead to injury to the walls of blood vessels during passage through the circulatory system. However, such a deleterious effect has not been demonstrated in human trials, and it is difficult to know the accuracy of extrapolating from rat to humans.

4.3.6.4 Phytosterols and Platelet Aggregation

In a mechanism similar to aspirin, consumption of phytosterol-enriched margarine can weaken the process involved in platelet adhesion and aggregation. This beneficial effect is also directly correlated with reduction in plasma atherogenic lipids. By prolonging platelet adhesion and aggregation time, phytosterols provide beneficial effects on blood flow and prevent formation of vascular blood clots that can be potentially fatal.

4.3.7 Proanthocyanidins (PAs)

These are polyphenols that contribute to the colors of several fruits and vegetables such as grapes, cherries, plums, blueberries, and cranberries. PAs are high molecular weight forms of epicatechin, one of the most abundant plant polyphenolic compounds. PAs are soluble in aqueous solutions and will yield anthocyanidins when heated in acidic media. One of the most abundant food sources of PAs is the cranberry fruit, which contains five major types:

(a) Epicatechin-$(4\beta \rightarrow 8)$-epicatechin
(b) Epicatechin-$(4\beta \rightarrow 8, \quad 2\beta \rightarrow O \rightarrow 7)$-epicatechin
(c) Epicatechin-$(4\beta \rightarrow 6)$-epicatechin-$(4\beta \rightarrow 8, 2\beta \rightarrow O \rightarrow 7)$-epicatechin
(d) Epicatechin-$(4\beta \rightarrow 8, \quad 2\beta \rightarrow O \rightarrow 7)$-epicatechin-$(4\beta \rightarrow 8)$-epicatechin
(e) Epicatechin-$(4\beta \rightarrow 8)$-epicatechin-$(4\beta \rightarrow 8, 2\beta \rightarrow O \rightarrow 7)$-epicatechin

From the above, it is evident that PAs contain 2 basic types of linkages between the epicatechin units:

- The $4\beta \rightarrow 8$ (B-type)
- The $4\beta \rightarrow 8$ and $2\beta \rightarrow O \rightarrow 7$ interflavonoid bonds (A-type)

The size of PAs increases as the plant ages, and older plants or seeds will contain higher amounts of large (tetramers and higher), insoluble polymers than younger plants. Apart from insolubility, large-size PAs may not be physiologically available to the consumer due to insufficient rate of absorption from the GIT. This has been confirmed in cell culture experiments that showed complete absorption of the monomer, dimer, and trimer PAs whereas the polymers could not enter the cells but became partially stuck on the cell surface. Polymeric PAs are fermented by colon microflora into low molecular weight phenols that may then be absorbed. Therefore, consumption of younger plant parts, especially the edible seeds that contain high levels of monomers, dimers, and trimers, will better enhance blood levels of PAs and could translate to physiological benefits. PAs have been examined for several potential health benefits, but recent evidence suggests that they can be used as antitumor agents to limit pathological intensity of cancer. For example, grape seed PAs are able to stimulate the immune system in experimental mice as evidenced by the increased stimulation of lymphocyte transformation and enhancement of the production of tumor necrosis factor-α. In combination with enhanced lysosomal enzyme activity and phagocytotic capacity of macrophages, the presence of grape seed PAs in the diet led to upregulation of the immune system, which limited growth of Sarcoma 180 tumor cells. Because of their inhibitory effects on digestive enzymes, directly (reduced catalysis) or indirectly (protein precipitation), high levels of dietary PAs may also contribute to reduced food digestion and nutrient bioavailability. By reducing nutrient digestion, PAs could contribute to reduced caloric intake and associated health benefits such as weight loss and blood glucose management. However, it should be noted that PAs can chelate iron and interfere with iron absorption; therefore, consumption of PA products should be minimized in people suffering from iron deficiency anemia. In rat experiments, addition of PAs to the diet of anemic rats led to toxic effects and even fatality.

4.3.7.1 Production of PAs Extracts

Cranberry PAs can be extracted by homogenizing the seeds in 95% aqueous ethanol that contains 1,000 ppm content of sulfur dioxide. The homogenate is then filtered and centrifuged to produce an ethanolic extract, followed by evaporation of the solvent and dissolving the residue in water. Pure aqueous extraction can also be used, which will remove mostly the soluble monomers, dimers, and trimers. Since other soluble colloidal polymers may also be extracted by the water, the extract can be subjected to microfiltration to separate the PAs from the unwanted polymers. The resultant dilute aqueous PAs extract can then be freeze-dried or concentrated into syrup, whichever method is most convenient. The solvent extract will contain most of the PAs, including the high molecular weight polymers. It should be noted that cranberry juice has reduced levels of bioactive PAs due to the negative effects (polymerization and degradation) of processing conditions and the fact that the seed skin does not become part of the pressed juice. Since the skin is very rich in PAs, total seed extraction is more effective than juice processing as a way of producing highly concentrated and bioactive cranberry product.

4.3.7.2 PAs and Urinary Tract Infection

The ability to prevent bacterial growth and multiplication in the urinary tract is one of the most demonstrated properties of food-derived PAs, especially from cranberry fruits. Urinary tract infection (UTI) refers to the infection of any or most parts of the urinary tract; 90% of all UTI is due to the *Escherichia coli* bacteria. *E. coli* is able to populate the urinary tract because of the presence of tiny proteinaceous hairy-like structures called fimbriae that enhance ability of the bacterial cells to cling onto the walls of the urinary tract. The ability to cling onto the urinary tract wall allows the bacterial to grow, multiply, and produce toxins that cause pathological symptoms in the host. Prevention of UTI by PAs is

because they also bind to the walls of the urinary tract and competitively prevent bacterial fimbriae from being able to bind. Unable to bind, the bacteria cells are washed away without being able to cause infection. There are two types of *E. coli* fimbriae, type 1 and type P, that are involved in UTI. Type 1 *E. coli* contain the mannose-sensitive fimbriae, which mean that they attach to the urinary tract wall through mannose residues that are present on the epithelial cells that line the tract. Type P fimbriae have no affinity for mannose, but instead they bind to the urinary tract through the glucose or galactose units on epithelial cells. Type P is referred to as the mannose-resistant fimbriae. Cranberry juice is believed to contain molecules that prevent binding of both types of *E. coli* fimbriae.

Firstly, the PAs prevent the type P fimbriae from adhering to the urinary tract, either through direct inhibition or indirectly through a genetic mode by interfering with ability of the *E. coli* to express the correct conformation or composition of fimbriae units.

Secondly, the free fructose molecules in cranberry juice are capable of replacing the normal mannose units on the urinary tract epithelial cells. Since the type 1 fimbriae have a strict requirement for mannose binding, the presence of fructose units prevents the *E. coli* from binding to the urinary tract lining. The actions of PAs and fructose lead to increased removal of the *E. coli* bacteria from the urinary tract as a result of weak binding or total inability to bind to the epithelial cells. In various human trials, it has been shown that daily consumption of cranberry juice can reduce UTI by as much as 14–20% and lead to a reduction in annual antibiotic consumption. Consumption of cranberry juice was found to be less cost-effective than the use of cranberry pills.

4.3.7.3 Cranberry Juice Polyphenols and Cancer Prevention

The role of cranberry juice extracts in cancer prevention has been investigated mostly using tissue culture techniques where there is direct contact with malignant cells. For example, a derivative of the B2 procyanidin dimer was found to exhibit cytotoxic activity against the human leukemic cell line HL-60 with an IC_{50} value of 119 µM that is similar to that of epigallocatechin gallate. The procyanidin dimer was also active against melanoma cell lines but inactive against several other tumor cell lines. Growth of human lung and colon carcinoma cell lines were inhibited by several galloylated PA dimers. In comparison, the non-galloylated dimers, A-type dimers, trimers, and a galloylated pentamer were not as active against cancer lines as the galloylated PA dimers. In separate experiments, it was shown that various PAs could inhibit the activity of protein kinase C (PKC), an enzyme that promotes signal transduction and tumor growth. Further proof was obtained when the PAs inhibited tumor growth that is caused by 12-O-tetradecanoylphorbol-13-acetate (TPA), which is a phorboid receptor for PKC. As a strong antioxidant, PAs may inhibit oxidative reactions that induce tumor formation, especially angiogenesis.

Inhibition of bacterial adhesion is a well-known mechanism through which PAs prevent infections, especially those caused by *Helicobacter pylori*. Because the incidence of gastric cancer has been shown to be positively related to *H. pylori* infection, dietary PAs may serve as anticancer agents by preventing adhesion of the bacterial cells to the GI tract, especially the stomach lining. In a clinical trial, the rate of infection by *H. pylori* was significantly reduced in adults that were fed cranberry juice when compared to the placebo group. In general, cranberry juice PAs and other flavonoids such as quercetin prevent the in vitro proliferation of cancer cells such as those from breast, colon, pancreas, and blood through various mechanisms that include:

- Induction of apoptosis
- Inhibition of epidermal growth factor receptor expression and associated tyrosine kinase activity
- Reduced expression of Ras protein
- Increased expression of endogenous inhibitors of matrix metalloproteinases
- Phytoestrogenic activity involving interactions with estrogen receptors in breast tissue
- Inhibition of ornithine decarboxylase expression

- Anti-inflammatory action such as inhibition of cyclooxygenase-2 expression

However, the therapeutic role of PAs against cancer remains to be demonstrated in animal or human experiments. Cranberry PAs can inhibit adhesion of cancer-promoting *Helicobacter pylori* cells to the gastric mucus and thereby prevent development of gastric tumors.

4.3.7.4 Cranberry Juice Polyphenols and Oral Health

Diseases of the oral cavity such as gingivitis, dental caries, and periodontitis may be prevented or treated with oral administration of cranberry juice. Bacterial cells that are responsible for oral cavity diseases will produce and become embedded within biofilms that coat hard and soft tissues. Formation of oral biofilms allow bacteria to populate the oral cavity, especially tooth surfaces where they produce various organic acids, especially lactic that ultimately leads to lowering of pH to less than 5.5. Cariogenic bacteria that are embedded in the oral biofilms are highly aciduric and acidogenic. Because the tooth enamel is highly susceptible to dissolution when the microenvironment becomes increasingly acidic, dental caries results from highly populated oral cavity biofilms. Cranberry extract and its PAs act as anticaries agents through four main types of properties:

(a) *Reduction in Hydrophobicity of Bacterial Cells.* Attachment of bacterial cells to tooth enamel surface is primarily through hydrophobic interactions. Hydrophilic PAs and other cranberry components reduce bacterial cell hydrophobicity and weaken the interactions between the bacteria and the tooth surface.

(b) *Reduction in Polysaccharide Synthesis.* Polysaccharides such as glucans and fructans act as anchor points on the tooth enamel and are synthesized mainly by two enzymes, glucosyltransferase (GTF) and fructosyltransferase (FTF). Cranberry extracts inhibit activities of GTF and FTF, which leads to reduced number of glucan anchor points on the tooth enamel.

(c) *Inhibition of Glucan Binding.* Cariogenic bacteria such as Streptococci interact with the tooth enamel through various polysac-charides using proteins that are present on the bacterial membrane. Consumption of cranberry juice helps to reduce capacity of bacterial cells to adhere to tooth surface, which reduces biofilm mass (thin and less dense) and leads to enhanced dental hygiene. This is due to interference from cranberry PAs that leads to reduced interactions between bacterial membrane proteins and tooth surface-bound polysaccharides.

(d) *Reduction in Acid Production.* Loss in tooth integrity is due mainly to increased acid pro-duction by the bacteria species such as the *Streptococci mutans* that are found in oral biofilms. Cranberry components, especially the PAs, prevent acid production by *S. mutans*, which ensures that enamel remains above the critical value of approx. pH 5.5 where demineralization is more likely to occur.

In periodontal diseases, the oral cavity becomes infected with a specific group of Gram-negative anaerobic bacteria, which leads to destruction of tooth-supporting tissues, especially the alveolar bone and the periodontal ligament. The adverse effects of these bacteria species are due mainly to production of proteolytic enzymes and the overproduction of inflammatory com-pounds such as proinflammatory cytokines and prostanoids in response to the presence of the periodontopathogens (e.g., *Porphyromonas gin-givalis, Fusobacterium nucleatum, Tannerella forsythia, Treponema denticola, Aggregatibacter actinomycetemcomitans*, and *Prevotella interme-dia*). Infection also stimulates the host to produce excess amounts of matrix metalloproteinases (MMP), which are critical factors in the initiation and progression of periodontitis. The potential ability of cranberry components to inhibit patho-gens that cause periodontitis and provide oral health benefits is manifested through three main mechanisms:

1. Inhibition of tooth adherence by periodontitis pathogens such as *P. gingivalis* and in this way prevents formation of viable oral biofilms. Co-aggregation of the bacterial cells is also inhibited, which prevents colonization of sub-

gingival sites. By preventing adhesion rather than microbial growth, the use of cranberry components as oral therapeutic agents will likely not result in the development of resis-tant bacterial strains.

2. Inhibition of proteases, especially the MMPs, which prevents destruction of bone and con-nective tissues.

3. Reduced host-mediated production of proinflammatory cytokines, prostaglandins, nitric oxide, and reactive oxygen species. Ultimately, there is reduction in the level of host immuno-inflammatory response.

4.3.8   Plant Anthocyanins

The purple color of certain varieties of corn grown in South America is due mainly to the presence of high contents of anthocyanins and other bioactive polyphenolic compounds. The major polypheno-lic pigment in purple corn is cyanidin 3-glucoside, which has been shown to be a potential chemo-preventive and chemotherapeutic agent. Recent evidence from human renal mesangial cell culture and mice experiments has shown that anthocya-nin-rich purple corn polyphenolic extracts can ameliorate diabetic nephropathy (DN), which is characterized by extracellular matrix (ECM) accumulation-dependent renal hardening and interstitial fibrosis. As diabetes pathology pro-gresses, there is development of nephrosclerosis, a condition characterized by thickening of glom-erular arteries because of accumulation of myofibroblasts and dense collagen tissues. DN leads to reduced functioning of the kidneys and occurrence of proteinuria, abnormal glomerular filtration, and alterations in secretion of renal bio-markers. This hyperglycemia-induced nephropa-thy is stimulated by increased activities of fibrogenic and prosclerotic cytokines (mainly TGF-β and MCP-1) that enhances ECM deposi-tion and inflammation. Treatment of human renal mesangial cells with the butanol extract of purple corn polyphenols attenuated the high glucose-induced cell proliferation and ECM accumula-tion. Specifically, the purple corn extract reduced the cellular expression of connective tissue growth factor (CTGF) by mesangial cells. CTGF contributes to increased fibrosis during the nephrosclerosis process by enhancing collagen accumulation; therefore, reduced activity can be beneficial for attenuating disease progression during DN. Thus, ability of purple corn anthocyanins to attenuate DN disease progression is linked to inhibition of IL-8 and hence CTGF productions.

During angiogenesis and cell proliferation, increased activity of platelet-derived growth factor (PDGF) plays a vital role. High glucose treatment of the mesangial cells led to increased activity of PDGF-BB (the dimeric form consisting of two B chains), which was inhibited when purple corn polyphenol extract was added. Also, addition of PDGF-BB to the cells triggered increased cell proliferation, which was attenuated in a dose-dependent manner by purple corn polyphenol extract. Therefore, it is plausible that ability of purple corn polyphenol extract to prevent mesangial hyperplasia-induced glomerulosclerosis is mediated through inhibition of cellular expression of PDGF-BB. Similarly, high glucose treatment of the mesangial cells led to upregulation of IL-8 secretion, which was also attenuated in the presence of purple corn polyphenol extract. IL-8 is a proinflammatory cytokine that enhances cellular production of CTGF and collagen IV, which indicates a role for this cytokine in ECM deposition during DN. High blood glucose as evident during diabetes leads to activation of the proinflammatory Janus kinase/signal transducers and activators of transcription (JAK/STAT) signaling pathway, which is also mediated by IL-8. For example, treatment of the mesangial cells with IL-8 produced phosphorylated tyrosine kinase 2 (Tyk2, the first member of the JAK family); this activation was also attenuated by addition of purple corn polyphenol extract. The purple corn polyphenol attenuation of Tyk2 activation led to reduced cellular expression of Tyk2-dependent proteins such as STAT1 and STAT3. Thus, by reducing activity of the mesangial inflammatory signaling pathway, the purple corn polyphenols were able to reduce intensity of diabetic renal fibrosis and glomerulosclerosis.

The purple corn polyphenol extract was also fed to diabetic (db/db) mice and shown to reduce plasma level of glycated hemoglobin, a marker for diabetic complications. Blood glucose was also significantly reduced by the polyphenol extract while plasma glucose was restored faster during oral glucose test when compared to the untreated diabetic mice. After an 8-week feeding experiment, the presence of purple corn polyphenol extract in the mice diet led to significant decreases in urinary albumin and creatinine levels. Histological examination of the kidneys showed reduced collagen staining in the purple corn extract-fed mice, which indicated alleviation of DN-associated glomerular fibrosis. The purple corn extract diet reduced cellular expression of collagen IV along with lower levels of TGF-β and CTGF in renal tissues. Analysis of the kidney for enzymes that play a role in ECM accumulation showed attenuated levels of tissue inhibitor of metalloproteinases (TIMP), which indicates reduced capacity for inhibiting matrix protein degradation when the mice diet contained the purple corn polyphenol extract. Reduced activity of TIMP will enhance protein degradation capacity in the kidney and prevent ECM accumulation. In contrast, presence of purple corn polyphenol extract in the mice diet led to upregulated levels of membrane type-1 matrix metalloproteinase (MT-1 MMP), an enzyme that can breakdown ECM. Combined with the decreased activity of TIMP, the high levels of MT-1 MMP observed during feeding of mice with purple corn extract-containing diet led to enhanced degradation of ECM, which ameliorated DN-induced glomerulosclerosis. In DN, there is severe decrease in renal levels of nephrin and podocin, which is an indication of a damaged renal filtration barrier. This is because nephrin is a required protein for adequate functioning of the renal filtration barrier. For example, a defect in the gene for nephrin is associated with congenital nephrotic syndrome in some populations, and it causes massive amounts of protein to be leaked into the urine, or proteinuria. Podocin is a protein component of podocytes, which is also part of the renal filtration barrier. Mutations in the podocin gene have been known to cause glomerulosclerosis and associated impaired renal functions. In mice that consumed purple corn extract-fortified diet, there were increased expression levels of nephrin and podocin in the kidneys, and glomerular staining suggested that damage to the impaired renal filtration barrier may actually have been repaired. This is because the level of glomerular staining in the purple corn extract-fed rats was similar to that of normal (db/m) mice that served as controls. Overall, the anthocyanin-rich purple corn extract produced anti-inflammatory, anti-glycative, and anti-glomerulosclerosis effects through downregulation of mesangial hyperplasia and repair of glomerular filtration barrier system. Therefore, the purple corn anthocyanin extract may be used to formulate functional foods and nutraceuticals aimed at preventing disease progression and reducing pathological intensity of DN.

An anthocyanin concentrate was also produced from purple sweet potato through ethanol/water (85:15, v/v) extraction and C-18 elution using acidified methanol as the eluting solvent. Evidence from rat and human studies has confirmed bioavailability of plant anthocyanins, with the metabolites shown to be present in urine after oral consumption. The in vivo antioxidant effects of the anthocyanins have also been demonstrated in rat-feeding experiments. An anthocyanin extract was fed to diet-induced obese mice by daily gavage for 4 weeks followed by testing for metabolic changes and potential effects against obesity-related pathological symptoms. Obese mice that received anthocyanin extract had significantly less body weight, liver weight, and serum hepatic enzymes as well as serum glucose when compared to the untreated mice. Since food intake did not differ between the anthocyanin-treated and nontreated mice, the data shows that anthocyanins could attenuate diet-induced body weight gain. The anthocyanin concentrate also reduced the high-fat diet-induced triglyceride accumulation in the liver as well as total hepatic cholesterol and lipid peroxidation. The decreased level of hepatic lipids associated with the sweet potato anthocyanin treatment was due to attenuated expression of sterol regulatory element-binding protein (SREBP) and its downstream target of fatty acid synthase, a lipogenic

enzyme. There was also evidence that the anthocyanin treatment upregulated lipid catabolism through increased expression of adenosine monophosphate-activated protein kinase (AMPK), an enzyme that decreases activity of ACC and switches off anabolic pathways. ACC catalyzes the biosynthesis of fatty acids and can also inhibit mitochondria fatty acid oxidation. Therefore, upregulated levels of AMPK led to decreased activities of ACC, which also favors increased activity of CPT-1. CPT-1 catalyzes translocation of long-chain fatty acyl-CoA into the mitochondria where they are oxidized. Overall, the purple sweet potato anthocyanins showed high potential use in decreasing pathological intensity of obesity-related symptoms through reduced lipid synthesis coupled with increased lipid catabolism and decreased blood sugar level.

4.3.9 Pomace Olive Oil Triterpenoids and Polyphenolic Constituents

Subsequent to the mechanical extraction of virgin olive oil, the residual pomace is further extracted to give "Orujo" oil that is rich in triterpenoids such as oleanolic and maslinic acids (Fig. 4.6), as well as the alcohols erythrodiol and uvaol.

Oleanolic acid is an effective antioxidant that protects LDL from oxidative damage in addition to other reported bioactive properties such anti-inflammatory, hepatoprotective, and antitumoral effects. Oleanolic acid also possesses vasodilation, antihyperlipidemic, and diuretic-natriuretic activities, which are useful for the prevention of hypertension. Therefore, pomace olive oil may serve as a functional food for the prevention of cardiovascular damage and as a hypotensive agent. In cell culture experiments, 95 and 28 µg/ml (noncytotoxic doses), respectively, of olive oil polyphenolic extract and gallic acid (present in olive oil) were able to reduce level of free radicals after oxidative stress induction with H_2O_2. The olive oil polyphenolic extract or gallic acid also attenuated oxidative stress-induced increases in caspase 9, a proapoptotic factor. Therefore, some of the protective effects of olive oil against

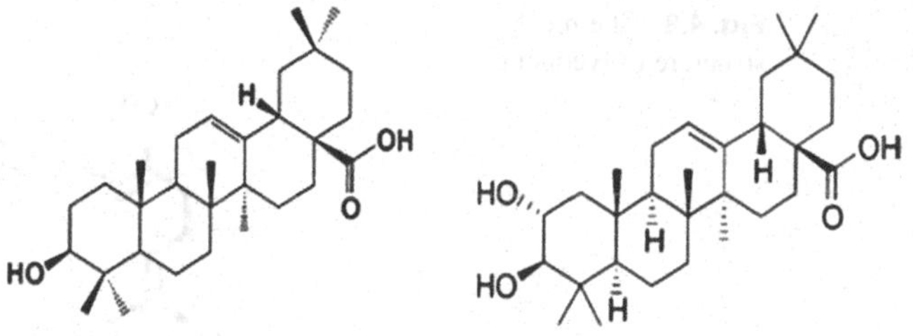

Fig. 4.6 Structure of olive oil triterpenoid compounds

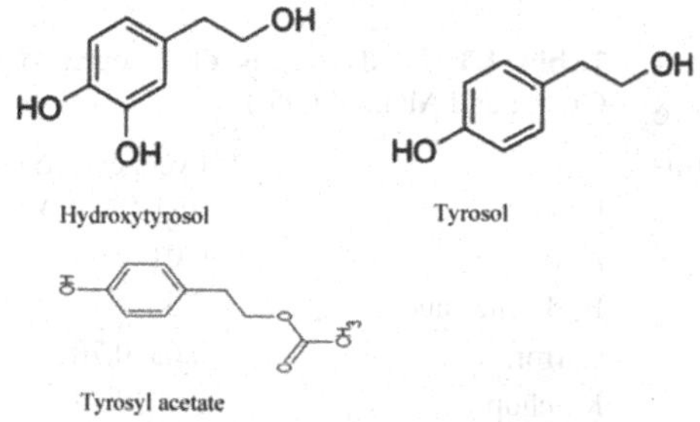

Oleanolic acid   Maslinic acid

Hydroxytyrosol   Tyrosol

Tyrosyl acetate

Fig. 4.7 Structures of olive oil phenolic compounds

oxidative stress could be linked to the antioxidant properties of the polyphenolic and triterpenoid constituents.

Olive oil also contains highly active phenolic compounds (Fig. 4.7) in the form of 2-(3,4-dihydroxyphenyl)ethanol also called hydroxytyrosol (HTy) and structural derivatives such as 2-(4-hydroxyphenyl)ethanol or tyrosol (Ty), and 2-(3,4-dihydroxyphenyl)ethyl acetate or hydroxytyrosyl acetate (HTyAc). In addition, there are other minor phenolic compounds such as 2-(4-hydroxyphenyl)ethyl acetate or tyrosyl acetate, phenolic derivatives of benzoic and cinnamic acids, vanillin, certain flavonoids, and lignans.

These phenolic constituents contribute to the antioxidant properties of olive oil with studies that have showed bioavailability of HTy and Ty. Once absorbed into the blood circulatory system, HTy and Ty are transported to the liver where they exert antioxidant and other biological effects before they are metabolized into glucuronidated and methylated conjugates for excretion in the urine. Apart from the demonstrated antioxidant properties (free radical scavenging, metal chela-

tion, and inhibition of lipid peroxidation) of olive oil phenolic extracts, the potential biological effects of HTy have also been studied. Initial in vitro evaluation with human hepatocarcinoma HepG2 cells showed that HTy was nontoxic and did not have any negative effect on cell structure. Cells that were treated with an oxidizing agent (tert-butyl hydroperoxide) and HTy maintained glutathione (GSH) concentrations that were similar to normal cells that received no hydroperoxide treatment. In contrast, cells treated only with the hydroperoxide but no HTy had significantly reduced levels of GSH, a very potent cellular antioxidant. Similarly, hydroperoxide treatment of cells led to significantly increased levels of glutathione peroxidase (GPx) and ROS as well as malondialdehyde, a lipid peroxidation product; pretreatment of cells with HTy suppressed these peroxide-induced degenerative effects. The low levels of GPx (catalyzes oxidation of GSH to GSSG) and the high levels of GSH during HTy treatment suggest reduced oxidative stress in the cells, which confirms the antioxidant ability of HTy. Therefore, olive oil or its main phenolic constituent, HTy, could be used as functional foods or in nutraceuticals forms to prevent pathogenesis of chronic degenerative diseases that are due to excess levels of tissue ROS.

4.4 Carotenoids

These are phytochemicals that are abundant in fruits and vegetables, and their increased consumption is associated with several health benefits. The carotenoids as a group consist of over 600 fat-soluble pigments that are responsible

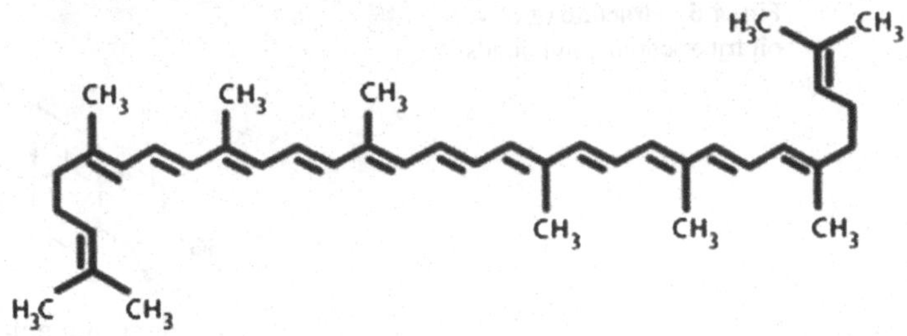

Fig. 4.8 Chemical
structure of lycopene

for the natural colors (yellow, orange, red) of fruits and vegetables. A major structural feature of carotenoids is the presence of several conjugated double bonds.

4.4.1 Lycopene

This is the principal carotenoid that is responsible for the distinctive red color of ripe tomato fruits (*Lycopersicon esculentum*) and tomato products. Apart from tomatoes, lycopene (Fig. 4.8) can also be found in watermelon, pink grapefruit, apricots, pink guava, and papaya, though tomato products (ketchup, tomato paste, pizza sauce, spaghetti sauce, tomato soup) remain the most abundant sources (Table 4.1).

Because humans cannot synthesize lycopene, food products are the only means of supplying them as potential bioactive compounds. Lycopene bioavailability is enhanced when present in the *cis*-configuration and when ingested with dietary fats. The preferential absorption of the *cis*-isomer when compared to the *trans*-isomer has been attributed to increased solubility of the *cis*-isomer in bile acid micelles, which enhances incorporation into chylomicrons. Food processing techniques such as heating or cooking enhance formation of the *cis*-isomer. In general, increased consumption of lycopene-containing food products is believed to have positive impact on human health and may actually prevent the occurrence of some chronic diseases. However, data from epidemiological studies is yet to provide a clear-cut role for lycopene in human health. Initial studies suggested that high consumption of tomatoes and tomato products may be associated with lower incidence of cardiovascular diseases and

Table 4.1 Food sources of lycopene (Adapted from Omoni and Aluko (2005))

Food type	Lycopene content (mg/100 g wet weight basis)
Apricot	0.01–0.05
Barbecue sauce	7.6
Carrot	0.65–0.78
Ketchup	9.0–13.44
Pizza sauce (canned)	12.71
Pink guava	5.23–5.50
Pumpkin	0.38–0.46
Tomatoes (fresh fruits)	0.72–20
Tomato paste	5.40–150
Tomato sauce	6.20
Spaghetti sauce	9.3–18.2
Sweet potato	0.02–0.11
Watermelon	2.3–7.2

could reduce the risk of cancers such as prostate, breast, lung, and digestive tract. Moreover, the risk of these chronic diseases was found to be inversely proportional to the amount of lycopene found in tissues and serum of human subjects. But recent epidemiological seem to suggest that consumption of tomato products does not protect humans against prostate cancer. In vitro tests have confirmed lycopene as a very potent anti-oxidant that could protect people from the damaging effects of reactive oxygen species and other endogenous free radicals. Therefore, most of the current evidence for the health benefits of lycopene has come from clinical studies as well as animal and tissue culture experiments, some of which are described below.

4.4.1.1 Lycopene and Cancer Prevention

In human endometrial, mammary, and lung cancer cells, lycopene was superior to α- and β-carotene in preventing tumor growth. For example, consumption of 10 or more weekly servings of tomato products was associated with up to 34% decrease in the risk of developing prostate cancer. Similarly, men with higher plasma levels of lycopene have a 25% reduction in overall prostate cancer risk and 44% reduction in the risk of aggressive prostate cancer. A plausible mechanism is that lycopene interferes with the production of insulin-like growth factor 1 (IGF-1), which is known to stimulate cancer cell growth and development. Lycopene-supplemented diet led to 29% reduction in plasma levels of IGF-1 in prostate cancer patients. It is believed that IGF-1 acts as a mitogen (stimulates cell division) for prostate cancer cells; thus, downregulation of IGF-1 by lycopene will reduce growth and proliferation of the cancer cells. Similar growth inhibition by lycopene has been demonstrated in prostate cancer cell culture models. In lycopene-fed mice, there was delayed onset and growth retardation of spontaneous mammary tumor when compared to the control mice that did not have lycopene in their diet. Inhibition of mammary cancer progression is associated with reduced activity of mammary gland thymidylate synthetase, an enzyme that is essential for the synthesis of DNA. Lycopene-fed mice had less amounts of serum free fatty acids and prolactin, a cell division-stimulating hormone that is implicated in breast cancer development. In multiorgan carcinogenesis mice model (B6C3F1), lycopene reduced lung cancer formation in a dose-dependent manner in males but not in females. As a preventive agent, lycopene is able to reduce formation of iron-induced formation of a DNA oxidation product (8-oxo-7,8-dihydro-2'-deoxyguanosine) in a cell culture model. By reducing DNA damage, lycopene helps to preserve the normal information encoded within genes and prevents development of abnormal cells such as those that cause tumors. The antioxidant effects of lycopene also lead to sparing of endogenous glutathione, which enhances hepatic glutathione-dependent detoxification of carcinogens.

In prostate cancer patients, shrinkage of tumors and reduced plasma level of prostate-specific antigen is associated with lycopene consumption. It has been suggested that daily consumption of 30 mg of lycopene could be used to modulate clinical markers of prostate cancer and may serve as a beneficial therapeutic intervention. Normal cells are known to communicate through gap junctions whereas tumor cells have dysfunctional gap junction communications (GJC). Lycopene stimulates proper functioning of GJC, which helps in preventing formation and progression of cancers. Gap junctions are water-filled pores that connect the cytosols of two neighboring cells and contain transmembrane proteins called connexins. Gap junctions function by allowing exchange of low molecular weight compounds between cells for normal cell growth. In tumors, the gap junctions are dysfunctional, which affects normal control of cellular exchanges and inability of the cells to function normally. For example, decreased expression of connexins in tumors has been reported and lycopene-supplemented diet led to increased level of connexin43 in the prostate gland of prostate cancer patients. Cell culture work with oral cavity cancer cells also showed increased upregulation of mRNA expression for connexin43. In androgen-dependent human prostate cancer cells (LNCaP), treatment with lycopene led to significant inhibition of cell proliferation. The mechanism involves activation (increase mRNA synthesis) of peroxisome proliferator-activated receptor gamma (PPARγ) and liver X receptor alpha (LXRα). The increased levels of PPPAγ and LXRα then lead to upregulated synthesis of ATP-binding cassette transporter 1(ABCA1), which mediates cholesterol efflux from cells and enhance reverse cholesterol transport from peripheral tissues into the liver. As cholesterol is removed from the cells, the membrane lipid raft domains (a platform for cellular signal transduction) become depleted of cholesterol, which reduces cell signaling and capacity for growth and proliferation. In fact, treatment with lycopene led to increased expression levels of ABCA1 and decreased intracellular cholesterol levels in the LNCaP cells, indicating that cholesterol efflux from the cells contributes to decreased survival of the cancer cells.

4.4.1.2 Lycopene in the Prevention of Atherosclerosis and Obesity

As previously stated in other sections of this book, oxidation of LDL molecules can induce development and progression of atherosclerosis, a proven marker for hypertension, heart attacks, and ischemic strokes. One of the proven activities of lycopene is as an antioxidant that prevents oxidative destruction of LDL, as shown in various in vitro and human subject experiments. Serum and tissue concentrations of lycopene are inversely related to the thickness of the innermost wall of blood vessels and the risk of myocardial infarction, which suggests an antiatherogenic role for lycopene. This is because a thick artery wall is a sign of early development of atherosclerosis while increased thickness of the intima media is a good predictor of coronary events. The presence of calcified plaques in the abdominal aorta (an indication of atherosclerosis) has been shown to be inversely correlated to serum lycopene concentrations, especially in smokers. In contrast, there was no risk reduction associated with serum levels of α-carotene, β-carotene, lutein, and zeaxanthin. Apart from reducing oxidative damage to LDL through its antioxidant properties, lycopene has been shown (in vitro) to inhibit cholesterol synthesis in macrophages through reductions in the activity of 3-hydroxy-3-methylglutaryl-CoA reductase, an enzyme that catalyzes the rate-determining step during cholesterol biosynthesis. However, it is impossible to conclude that lycopene reduces cholesterol synthesis in living tissues, since the in vivo effect of lycopene on cholesterol biosynthesis has not been proven. Another important role for lycopene is that it can prevent adhesion of molecules on vascular endothelial cells, an important event during pathogenesis of atherosclerosis. When compared to other carotenoids, lycopene is the only compound that has been shown to reduce endothelial expression of adhesion molecules on cell surfaces, which enhances flow of materials through blood vessels by preventing deposition of materials such as LDL cholesterol. Lycopene consumption can increase plasma HDL-cholesterol, which could be one of the ways by which the antiatherogenic effects are manifested. This is because HDL is responsible for transportation of cholesterol

from the cells to the liver where they are excreted into bile and helps lower blood cholesterol level.

Lycopene can also prevent increases in total body, white adipose tissue, and liver weights that are associated with consumption of a high-fat diet, which suggests potential antiobesity effects. The mechanism is partly through inhibition of fatty acid synthesis and reduced lipid droplet size formation in white adipocytes. Specifically, activity of glucose-6-phosphate dehydrogenase (provides NADPH required for fatty acid synthesis) was reduced when diet of high-fat-fed rats was fortified with lycopene. Activity of phosphatidate phosphohydrolase, the enzyme that catalyzes the rate-limiting step during triglyceride synthesis, was also reduced in rats fed lycopene-supplemented diet. The antiobesity action of lycopene could also be due to its food efficiency lowering ability, which reduces nutrient absorption or metabolism. Lycopene suppresses accumulation of visceral fat depots such as the epididymal and perirenal adipose depots as part of the body weight-reducing effects. In contrast, brown adipose tissue (BAT) weight was increased by lycopene supplementation. BAT is known to play a role in increased thermogenesis, which promotes lipid oxidation rather than lipid storage. CPT-1, the rate-limiting enzyme in the transfer of long-chain fatty acids across the mitochondria membrane, was upregulated in lycopene-fed rats, which supports increased thermogenesis and lipid oxidation. Lycopene reduced development of fatty liver as evidenced by the lower triglyceride levels, which may be due to reduced absorption and enhanced fecal secretion. Glutamic oxaloacetic transaminase, alkaline phosphatase, and glutamate pyruvate are markers of liver damage; their levels were significantly reduced by lycopene in comparison to the control.

Type 1 diabetes is associated with impairment of endothelial function partly because of decrease NO levels in the aorta arising from decreased expression of endothelial nitric oxide synthase and scavenging of NO by free radicals (NO reacts with superoxide to form peroxynitrite). Additionally, there is high oxidative stress (increased plasma level of malondialdehyde, MDA) as well as decreased activity of antioxidant enzymes, especially superoxide dismutase (SOD).

Lycopene in the diet led to significant improvements in endothelial function of streptozotocin-induced diabetic rats as measured by decreased levels of MDA and increased level of SOD in the thoracic aorta. Plasma levels of oxidized LDL were also reduced by the lycopene diet, which prevents macrophage accumulation in the arterial wall. Blood glucose was significantly reduced in the diabetic rats by the lycopene diet, probably as a result of increased stimulation of insulin secretion, increased repair or regeneration of pancreatic β-cells, and enhanced antioxidant capacity.

Bibliography

AbuMweis, S.S., C.A. Vanstone, A.H. Lichtenstein, and P.J.H. Jones. 2009. Plant sterol consumption frequency affects plasma lipid levels and cholesterol kinetics in humans. *European Journal of Clinical Nutrition* 63: 747–755.

Al-Awwadi, N.A., A. Bornet, J. Azay, C. Araiz, S. Delbosc, J.-P. Cristol, N. Linck, G. Cros, and P.-T. Teissedre. 2004. Red wine polyphenols alone or in association with ethanol prevent hypertension, cardiac hypertrophy, and production of reactive oxygen species in the insulin-resistant fructose-fed rat. *Journal of Agricultural and Food Chemistry* 52: 5593–5597.

Al-Awwadi, N.A., C. Araiz, A. Bornet, S. Delbosc, J.-P. Cristol, N. Linck, J. Azay, P.-T. Teissedre, and G. Cros. 2005. Extracts enriched in different polyphenolic families normalize increased cardiac NADPH oxidase expression while having differential effects on insulin resistance, hypertension, and cardiac hypertrophy in high-fructose-fed rats. *Journal of Agricultural and Food Chemistry* 53: 151–157.

Biedrzycka, E., and R. Amarowicz. 2008. Diet and health: apple polyphenols as antioxidants. *Food Reviews International* 24: 235–251.

Bodet, C., D. Grenier, F. Chandad, I. Ofek, D. Steinberg, and E.I. Weiss. 2008. Potential oral health benefits of cranberry. *Critical Reviews in Food Science and Nutrition* 48: 672–680.

Bravo, L. 1998. Polyphenols: chemistry, dietary sources, metabolism, and nutritional significance. *Nutrition Reviews* 56: 317–333.

Cazzola, R., and B. Cestaro. 2011. Red wine polyphenols protect n-3 more than n-6 polyunsaturated fatty acid from lipid peroxidation. *Food Research International* 44: 3065–3071.

de Jong, A., J. Plat, and P. Mensink. 2003. Metabolic effects of plant sterols and stanols. *The Journal of Nutritional Biochemistry* 14: 362–369.

Garcia-Alonso, J., G. Ros, and M.J. Periago. 2006. Antiproliferative and cytoprotective activities of a phenolic-rich juice in HepG2 cells. *Food Research International* 39: 982–991.

Goya, L., R. Mateos, and L. Bravo. 2007. Effect of the olive oil phenol hydroxytyrosol on human hepatoma HepG2 cells. *European Journal of Nutrition* 46: 70–78.

Hope, C., K. Planutis, M. Planutiene, M.P. Moyer, K.S. Johal, J. Woo, C. Santoso, J.A. Hanson, and R.F. Holcombe. 2008. Low concentrations of resveratrol inhibit Wnt signal throughput in colon-derived cells: implications for colon cancer prevention. *Molecular Nutrition & Food Research* 52: S52–S61.

Hwang, Y.P., J.H. Choi, E.H. Han, H.G. Kim, J.-H. Wee, K. Ok, K.H. Jung, K.-I. Kwon, T.C. Jeong, Y.C. Chung, and H.G. Jeong. 2011. Purple sweet potato anthocyanins attenuate hepatic lipid accumulation through activating adenosine monophosphate-activated protein kinase in human HepG2 cells and obese mice. *Nutrition Research* 31: 896–906.

Jaffri, J.M., S. Mohamed, I.N. Ahmad, N.M. Mustapha, Y.A. Manap, and N. Rohimi. 2011. Effects of catechin-rich oil palm leaf extract on normal and hypertensive rats' kidney and liver. *Food Chemistry* 128: 433–441.

Kim, A.-Y., Y.-J. Jeong, Y.B. Park, M.-K. Lee, S.-M. Jeon, R.A. McGregor, and M.-S. Choi. 2012. Dose dependent effects of lycopene enriched tomato-wine on liver and adipose tissue in high-fat diet fed rats. *Food Chemistry* 130: 42–48.

Kozlowska-Wojciechowska, M., M. Jastrzebska, M. Naruszewicz, and A. Foltynska. 2003. Impact of margarine enriched with plant sterols on blood lipids, platelet function, and fibrinogen level in young men. *Metabolism* 52: 1373–1378.

Li, J., M.-K. Kang, J.-K. Kim, J.-L. Kim, S.-W. Kang, S.S. Lim, and Y.-H. Kang. 2012. Purple corn anthocyanins retard diabetes-associated glomerulosclerosis in mesangial cells and db/db mice. *European Journal of Nutrition*. doi:10.1007/s00394-011.0274-4.

Meydani, M., and S.T. Hasan. 2010. Dietary polyphenols and obesity. *Nutrients* 2: 737–751.

Moghadasian, M.H., D.V. Godin, B.M. McManus, and J.J. Frohlich. 1999. Lack of regression of atherosclerotic lesions in phytosterol-treated apoE-deficient mice. *Life Sciences* 64: 1029–1036.

Neto, C.C. 2007. Cranberry and its phytochemicals: a review of in vitro anticancer studies. *The Journal of Nutrition* 137: 186S–193S.

Nichenametla, S.N., T.G. Taruscio, D.L. Barney, and J.H. Exon. 2006. A review of the effects and mechanisms of polyphenolics in cancer. *Critical Reviews in Food Science and Nutrition* 46: 161–183.

Omoni, A.O., and R.E. Aluko. 2005. The anti-carcinogenic and anti-atherogenic effects of lycopene: a review. *Trends in Food Science and Technology* 16: 344–350.

Rodriguez-Ramiro, I., S. Ramos, L. Bravo, L. Goya, and M.A. Martin. 2012. Procyanidin B2 induces Nrf2 translocation and glutathione S transferase P1 expression via ERKs and p38-MAPK pathways and protect human colonic cells against oxidative stress. *European Journal of Nutrition.* doi:10.1007/s00394-011-0269-1.

Rodriguez-Rodriguez, R., J.S. Perona, M.D. Herrera, and V. Ruiz-Gutierrez. 2006. Triterpenic compounds from "orujo" olive oil elicit vasorelaxation in aorta from spontaneously hypertensive rats. *Journal of Agricultural and Food Chemistry* 54: 2096–2102.

Santos-Buelga, C., and A. Scalbert. 2000. Proanthocyanidins and tannin-like compounds- nature, occurrence, dietary intake and effects on nutrition and health. *Journal of the Science of Food and Agriculture* 80: 1094–1117.

Somova, L.I., A. Nadar, P. Rammanan, and F.O. Shode. 2003. Cardiovascular, antihyperlipidemic and anti-oxidant effects of oleanolic and ursolic acids in experimental hypertension. *Phytomedicine* 10: 115–121.

Tong, H., X. Song, X. Sun, G. Sun, and F. Du. 2011. Immunomodulatory and antitumor activities of grape seed proanthocyanins. *Journal of Agricultural and Food Chemistry* 59: 11543–11547.

Udenigwe, C.C., V.R. Ramprasath, R.E. Aluko, and P.J.H. Jones. 2008. Potential of resveratrol in anticancer and anti-inflammatory therapy. *Nutrition Reviews* 66: 445–454.

Wang, W., Y. Yagiz, T.J. Buran, do C. Nascimento, and L. Gu. 2011. Phytochemicals from berries and grapes inhibited the formation of advanced glycation end-products by scavenging reactive carbonyls. *Food Research International* 44: 2666–2673.

Yang, C.-M., I.-H. Lu, H.-Y. Chen, and M.-L. Hu. 2012. Lycopene inhibits proliferation of androgen-dependent human prostate tumor cells through activation of PPARγ-LXRα-ABCA1 pathway. *The Journal of Nutritional Biochemistry* 23: 8–17.

Zafriri, D., I. Ofek, R. Adar, M. Pocino, and N. Sharon. 1989. Inhibitory activity of cranberry juice on adherence of type 1 and type P fimbriated *Escherichia coli* to eukaryotic cells. *Antimicrobial Agents and Chemotherapy* 33: 92–98.

Zhang, L., J. Ma, K. Pan, V.L. Go, J. Chen, and W.-C. You. 2005. Efficacy of cranberry juice on *Helicobacter pylori* infection: a double blind, randomized placebo-controlled trial. *Helicobacter* 10: 139–145.

Zirk, M.M., R.E. Aluko, and C.G. Taylor. 2004. Cranberry (*Vaccinium macrocarpon*) proanthocyanidins and their effects on urinary tract infections. *Current Topics in Nutraceutical Research* 2: 153–161.

Part II

Specific Functional Foods

5.1 Introduction

While there have been recent doubts about the efficacy of soybean products as health-promoting products, the preponderance of evidence suggests that several health benefits are in fact associated with soybean products consumption. This chapter will shed some light on current knowledge on the use of soybean products as functional foods and nutraceuticals.

5.2 Bioactive Components

Soybean seeds are sources of several bioactive compounds such as the isoflavones that act as phytoestrogens because of structural and functional similarities to estradiol (Fig. 5.1). The main soybean isoflavones are the daidzein and genistein molecules which have demonstrated weak estrogenic effects. It should be noted that these phytoestrogens can have both estrogenic and antiestrogenic properties because they compete with estradiol to bind with estrogen receptors. However, phytoestrogens elicit much weaker estrogenic responses upon binding to estrogen receptors when compared to estrogen.

Soybean seeds also contain proteins that have been implicated in the improvements in cardiovascular health of regular consumers of soybean products. The effect of soybean proteins is associated with reductions in blood cholesterol through two main mechanisms:

1. Physical entrapment of bile acids in the gut by indigestible soybean protein fractions, which reduces cholesterol reabsorption in the colon and enhances their elimination in the feces.
2. It has also been shown in several experiments that digestion of soybean proteins releases peptides that inhibit some key enzymes involved in hepatic synthesis of cholesterol.

In addition to phytoestrogens and proteins, the high contents of polyunsaturated fatty acids in soybean oil and fiber in the seed coat contribute to the health benefits of soybean seeds.

5.3 Role of Soybean Components in Specific Disease Conditions

While the mechanisms for some of the health benefits of soybean seeds have not been fully elucidated, there is abundance of evidence from human and animal experiments that show that regular consumption of soybean foods can provide health benefits. In addition, epidemiological studies have shown lower incidences of heart disease, some forms of cancer, and menopausal symptoms in China and Japan where per capita consumption of soybean products is highest in the world. The following sections provide insight on the potential effects of soybean products in health and disease conditions.

R.E. Aluko, *Functional Foods and Nutraceuticals*, Food Science Text Series,
DOI 10.1007/978-1-4614-3480-1_5, © Springer Science+Business Media, LLC 2012

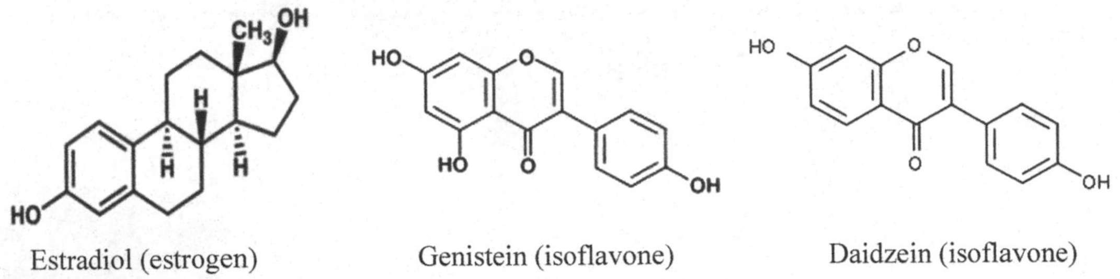

Estradiol (estrogen)   Genistein (isoflavone)   Daidzein (isoflavone)

Fig. 5.1 Chemical structures of estradiol and soybean isoflavones

5.3.1 Cardiovascular Diseases (CVD)

Soybean isoflavones may provide cardiovascular health benefits by improving arterial compliance (elasticity), a vascular function that normally decreases with age. Improved arterial function can enhance reduction of blood pressure and prevent development of high blood pressure-related diseases. However, the major beneficial effect of soybean in CVD is associated with its ability to reduce serum cholesterol, reduce lipid oxidation, and improve lipid profile in animal and human studies. That the beneficial effect was due to the intact protein was obvious when a mixture of amino acids had no cholesterol-lowering effects unlike the soybean protein. An approved health claim in the USA suggests that dietary soybean of 25 g or more per day (as part of a diet low in saturated fat) is associated with modest reductions in total and low-density lipoprotein (LDL) cholesterol levels, especially in people who already have elevated cholesterol levels. Soybean protein induces increased expression of superoxide dismutase (SOD), an antioxidant enzyme that converts the highly reactive and toxic superoxide anion into harmless water. Thus, by reducing the level of free radicals, soybean protein may prevent oxidative damage to LDL particles and prevent formation of foam cells that can block blood vessels. Soybean protein is believed to possess cholesterol-lowering ability because of its ability to upregulate LDL receptor levels in the liver, which increases LDL metabolism but decreases plasma concentration of LDL. Dietary soybean proteins also have beneficial effects in reducing body weight and fat mass. The body weight loss and hypocholesterolemic effects of soybean pro-teins have been confirmed in several human studies with statistically significant reductions of 6–12% in total plasma cholesterol. The body weight loss is due in part to the fact that soybean protein increases satiety and fat oxidation, which contributes to reduced postprandial plasma lipid and insulin contents. The increased satiety reduces overall caloric intake while high rate of fat oxidation reduces availability of fatty acid substrates that can be incorporated into adipose tissues. Apart from the beneficial effects of total or sole soybean consumption, its use as a partial replacement of animal proteins also reduced plasma cholesterol in hypercholesterolemic patients. The beneficial heart health effects of soybean proteins have been traced to the α′ sub-unit of the 7S globulin, which was shown to upregulate liver LDL receptor activity in mice with concomitant reductions in plasma cholesterol and triglyceride levels. In fact, feeding of a soybean variety that is devoid of the 7S α′ sub-unit was ineffective for cholesterol reduction. However, the results could not be replicated in primates, which cast doubt on the cholesterol-lowering efficacy of the soybean 7S globulin. But human intervention trials using soybean 7S globulin (5 g/day) have been shown to reduce body weight, triglycerides, and visceral fat while it increased HDL cholesterol in moderately over-weight Japanese individuals with hypertriglycer-idemia. Proposed mechanisms of action of soybean proteins include inhibition of fatty acid synthesis through downregulation of fatty acid synthase as well as reduction in cholesterol synthesis through inhibition of hydroxymethylglu-taryl-coenzyme A (HMG-CoA) reductase activity. During digestion in the gastrointestinal tract, soybean proteins also release bioactive peptides. The bioactive peptides are small enough to be absorbed into the blood circulatory system and carried into the liver where they inhibit enzymes involved in cholesterol synthesis or activate LDL receptors. It has also been shown that soybean 7S proteins may act as an antioxidant as evidenced by the fact that soybean diet increased the lag phase for copper-induced LDL oxidation when compared to casein diet. Apart from the 7S and 11S globulins, soybean seed also contains lunasin, a polypeptide that contains 43 amino acids (MW 5.4 kDa). The presence of lunasin in human plasma has been confirmed, indicating that absorption of the protein from soybean-containing diets is possible. Lunasin has been shown to exert hypocholesterolemic activity by blocking acetylation of histone H3 Lys14 residue, thereby reducing the production of HMG-CoA reductase with concomitant decrease in cholesterol biosynthesis. Lunasin also acts as a hypocholesterolemic agent by increasing cellular production of LDL receptors, which facilitates removal of LDL cholesterol from the plasma.

Interpretation of soybean protein's hypocholesterolemic effects can be confounded by the presence of isoflavones that are also known to have bioactive properties. In this aspect, no concrete conclusion has been made due to conflicting results from animal and human experiments. For example, there was no effect on cholesterol level when primates were fed casein diet that has been fortified with isoflavone-rich soybean extract. In contrast, premenopausal women who were fed a high-soybean isoflavone diet (128.7 mg/day) for three menstrual cycles had 7–10% reductions in LDL cholesterol. Similarly, consumption of soybean beverages by elderly men led to correlations between plasma cholesterol and total isoflavone concentration. In several human-feeding trials, consumption of isoflavones led to significant reductions in total and LDL cholesterol with concomitant increases in HDL cholesterol. In contrast, consumption of soybean proteins containing low (0.15 mg/g) isoflavone content had no beneficial effects on serum cholesterol reduction. Human trials and animal experiments have showed that a high-isoflavone diet increased

resistance of LDL to Cu^{2+}-induced oxidation in human subjects when compared to a low-isoflavone diet. Also important is the fact that the presence of isoflavones seems to spare depletion of other antioxidants because serum levels of tocopherol have been found to be increased after consumption of isoflavone-containing soybean diet. It is possible that differences in the dietary consumption of total amount of isoflavone and the ratio of the 12 major soybean isoflavones may be responsible for some of the conflicting results on the health benefits of soybean protein products. This is because the individual potency of the isoflavones may differ as well as the possibility of synergistic or antagonistic interactions within the isoflavone group or between the isoflavones and other phytoestrogens that may be present in a typical human diet. There is also the possibility that the rate and level of conversion of isoflavones to equol in the colon differs within the human population. Equol is known to induce increase production of nitric oxide (a vasodilator), which can prevent elevation of blood pressure. Therefore, people who metabolize isoflavone to equol at high rate will show better improvements in CVD symptoms than those that have low rate of isoflavone conversion. However, other available evidence suggests that daily consumption of soybean products (including isoflavones) may not have significant effects on cholesterol levels or reduce the risk of heart disease. Increased consumption of soybean products is still highly encouraged by the American Heart Association because soybean seeds contain higher levels of polyunsaturated fatty acids than meat or dairy products. Soybean seeds are also good sources of dietary vitamins, minerals, and fiber.

5.3.2 Renal Diseases

Dietary soybean protein has been tested in various animal and human studies as a means of ameliorating chronic kidney diseases. One of the leading causes of kidney diseases (nephropathy) is type II diabetes, and the associated high oxidative condition increases LDL cholesterol oxidation into reactive oxygen species that promote glomerular injury. In addition, type II diabetes is associated with high blood lipid profile that increases the risk of hypertension, which contributes to deterioration of kidney functions. Soybean protein has demonstrated antioxidant effects and blood lipid-reducing effects that can be used to modify the risk factors (high blood lipids and LDL oxidation) associated with nephropathy. In type II patients with nephropathy, 50% substitution of the dietary animal protein with soybean protein led to significant improvements in kidney functions measured as decreases in proteinuria, urinary urea nitrogen, and phosphorus. The soybean protein-substituted diet also led to significant reductions in plasma total cholesterol, triglycerides, and LDL cholesterol when compared to the animal protein diet. Importantly, the soybean-substituted diet did not affect HDL cholesterol, which is the beneficial form of lipoproteins. However, the soybean-substituted diet did not reduce hypertension (a major cause of proteinuria), which suggests that the reduction in proteinuria may be due to the ability of soybean proteins to reduce hyperfiltration that precedes onset of proteinuria in diabetes patients.

Soybean protein-supplemented diets have also been shown to be effective in modulating growth and disease progression of polycystic kidney disease. In contrast to casein-based diets, soybean protein-based diets have been shown to prevent cardiac hypertrophy and normalize serum creatinine levels with significant reductions in kidney weight, fibrosis, glomerular size, water content, and cyst size in Han:SPRD-*cy* rats, an experimental model of human chronic kidney disease (CKD). Kidney hypertrophy and fibrosis are part of the disease markers of CKD, and a role for insulin-like growth factor 1 (IGF-1) has been suggested because of significantly high concentrations in Han:SPRD-*cy* rats when compared to normal rats. Addition of soybean to the diet reduced IGF-1 levels in the kidney and plasma, when compared to a casein-based diet. Increase in kidney glomerular size is an early sign of kidney disease and is an independent risk factor for kidney disease progression. The ability of soybean protein to reduce glomerular size may be mediated through reduced production of 6-keto PGF1α, an eicosanoid that induce glomerular dilatation.

Soybean protein also reduced protein expression levels for cytosolic phospholipase A2 (cPLA$_2$) and COX-2 (but not COX-1) in addition to attenuated levels of their catalytic prostanoid products, thromboxane B2 (TXB$_2$), and prostaglandin E2 (PGE$_2$). However, overall, there were reduced ratios of TXB$_2$/PGE$_2$ and TXB$_2$/6-keto PGF1α, suggesting a modification of renal prostanoids in favor of vasodilation (PGE$_2$ and 6-keto PGF1α are vasodilatory agents). cPLA$_2$ catalyzes release of fatty acids from phospholipids; the fatty acids are substrates for COX-catalyzed reactions that lead to prostanoid production. Therefore, the decreased level of cPLA$_2$ suggests reduced substrate for COX, which resulted in decreased levels of prostanoids. Thus, COX-2 inhibition seems to be a potential way for inhibiting CKD disease progression. The increased risk of adverse cardiovascular events associated with COX-2 inhibitors was not observed in the soybean-fed rats. In fact, in contrast to casein diet, the soybean diet prevented Han:SPRD-*cy* rat cardiac hypertrophy that is normally associated with CKD. Thus, soybean protein may be a suitable alternative to drugs for the purpose of use as a COX-2 inhibitor in the management of CKD. Initially, the renoprotective effects of soybean were thought to be due in part to the presence of isoflavones, which could mimic hormones. However, the use of isoflavone-depleted soybean protein (0.11 mg isoflavone/g protein) produced similar renoprotective effects as the regular soybean protein (3.62 mg isoflavone/g protein), suggesting that proteins were mostly responsible for the observed bioactive properties. In fact, the isoflavone-depleted protein produced greater reductions in kidney weight, cell proliferation, and cyst size than the regular isoflavone-rich soybean protein. However, the reduction in level of oxidized LDL was less by the isoflavone-depleted protein, possibly because the higher isoflavone content in the regular protein enhanced greater antioxidant capacity.

5.3.3 Cancer

Epidemiological studies have shown lower levels of cancers, especially breast and prostate cancers in China and Japan when compared to Western countries. This trend is believed to be due to differences in diet and lifestyle, especially soybean food consumption that is higher in China and Japan. Therefore, it has long been suspected that soybean may have chemoprotective properties against certain forms of cancer. For example, Asian-American women that were born and raised in Western countries have a 60% higher risk of developing breast cancer when compared to those born in China and Japan. In various studies that followed the lifestyles of migrants, it was evident that those on Western-style diets (contains high fat and low fiber) have higher rates of breast cancer than those on traditional Eastern-style diets that is low in fat and relatively high in fiber and soybean foods. In human trials, there have been conflicting results with majority showing positive relationships between soybean consumption and reduced incidence of breast cancer. While some studies have implicated soybean isoflavones specifically in reduced rate of breast cancer development, others have shown no conclusive evidence that implicates consumption of phytoestrogens in reduced risk of breast cancer formation. Therefore, it has been suggested that the reduced risk of breast cancer development seen in epidemiological studies may be due to exposure to phytoestrogens during development or early life. In animal studies, administration of genistein had protective effects against chemically induced cancer but not against implanted mammary tumors, which again demonstrates the benefits of early life exposure to phytoestrogens. It is possible that the phytoestrogens could act as preventive agents but become ineffective once the tumors have established. Plausible mechanisms include the observations that genistein modulates cellular-signaling pathways and estrogen-responsive genes, interacts with the cell cycle, and alters cell differentiation. Using cell culture techniques, it was shown that a soy isoflavone preparation, containing mainly daidzein, genistein, and glycitein, effectively inhibited growth of cervical cancer cells. The isoflavone preparation was more effective than genistein in inhibiting growth of the cervical cancer cells, which suggests possible synergistic effect among the three compounds. The mechanism involved induction of apoptosis through the mitochondrial

pathway and regulation of nuclear factor-kappa B (NF-κB) signaling. Normal homeostasis requires programmed cell death (apoptosis) as a way of removing precancerous or cancerous cells. But apoptosis is known to be inhibited by NF-κB, which binds to DNA, causes transcription of genes such as Bcl-2 (translated into an antiapoptotic protein), and leads to decrease rate of apoptosis. When tested in various cancer cells, genistein effectively inhibited H_2O_2 or tumor necrosis factor alpha (TNFα)-induced NF-κB activation as well as the DNA-binding activity (through reduced translocation of NF-κB into the nucleus). Moreover, the soybean isoflavones reduced translocation of NF-κB. In human subjects, lymphocytes were collected after 3 weeks of consumption of soybean isoflavone supplements that contained genistein, daidzein, and glycitein. The lymphocytes were then treated with TNFα followed by analysis of NF-κB production. Lymphocytes from subjects that consumed the soybean isoflavone supplement did not show any activity of NF-κB, whereas in placebo subjects, the lymphocytes showed activated binding of NF-κB to DNA. Isoflavones have also been shown to induce apoptosis through regulation of other signaling pathways such as the Akt (protein kinase B, PKB), MAPK (mitogen-activated protein kinase), Wnt, Notch, p53, and AR (androgen receptor). Soy isoflavones also act as anticancer agents by inducing phase II enzymes such as glutathione-s-transferase and quinone reductase that enhance the body's ability to detoxify and excrete toxic products from metabolism of aromatic hydrocarbons. As antioxidants, isoflavones can reduce lipid peroxidation and prevent oxidative damage to DNA; 5-hydroxymethyl-2-deoxyuridine (a biomarker of DNA damage) was reduced in women that consumed 50 mg isoflavone tablets daily for 3 weeks. Overall mechanisms of action of the anticancer effects of isoflavones involve antioxidative effects as well as induction of apoptosis and inhibition of cellular proliferation.

In men, soybean isoflavones could have a role in the management and prevention of prostate cancer. Prostate cancer killed almost 30,000 men in the United States in 2005, second only to skin cancer in fatality. Prostate cancer is an androgen-dependent excessive cell proliferation, and treatment usually involves the use of compounds that have androgen-deprivation properties. But as the disease advances in pathology, the cancerous cells become resistant to androgen-deprivation therapy, which leads to metastasis and eventual death. Fundamentally, the prostaglandin (PG) pathway contributes to the severity of prostate cancer because prostaglandins promote tumor growth and metastasis. Therefore, compounds that modulate the PG pathway are important candidates for the treatment of prostate cancer. The PG pathway and metabolic action are heavily dependent on the activities cyclooxygenases (COX), especially COX-2 as well as PG receptors. In contrast, inactivation of PG is heavily dependent on the activity of 15-hydroxyprosta-glandin dehydrogenase (15-PGDH); therefore, 15-PGDH is a tumor-suppressing factor. Reduced expression of 15-PGDH contributes to excessive levels of prostaglandins, especially PGE_2, the most significant PG produced in human prostate tissue. Genistein acts against prostate cancer by reducing production of PGE_2 through suppression of COX-2 activity and by enhancing the ability of 1,25-dihydroxycholecalciferol (the active form of vitamin D found in the body) to induce high cellular levels of 15-PGDH. Genistein also acts by inhibiting PG receptors which prevents cellular uptake of PG, especially PGE_2 in prostate tissue. Therefore, by inhibiting the synthesis and biological effects of PGs, soybean genistein has a role in the prevention and therapeutic treatment of prostate cancer. However, it should be emphasized that excessive consumption of isoflavone-rich foods could have negative effects such as hypogonadism (decreased testosterone production) and erectile dysfunction in men. While the exact upper intake level of isoflavone is not fully understood, recent report showed that a man with average consumption of 360 mg isoflavone per day over a 1-year period suffered from complete loss of libido and erectile dysfunction. The patient had decreased percentage of free plasma testosterone in addition to increased level of dehydroepiandrosterone (DHEA); these symptoms were mostly eliminated after patient discontinued a soybean-laden

vegan diet. It was speculated that the high content of dietary isoflavones blunted testosterone formation by reducing DHEA conversion to testosterone through the activities of genistein and daidzein, inhibitors of DHEA-converting enzymes (3-β-hydroxysteroid dehydrogenase and 17-β-hydroxysteroid dehydrogenase). Another possible explanation for the negative effects of high isoflavone consumption is its ability to enhance synthesis of sex hormone-binding globulin, which binds testosterone and leads to decreased level of circulating free testosterone.

Apart from the isoflavones, the presence of the Bowman-Birk protease inhibitor (BBI) may also contribute to the anticarcinogenic properties of soybean protein products. BBI acts as an anticancer agent by:

1. Blocking stimulated neutrophil-induced formation of active oxygen species
2. Inhibiting tumor promotion
3. Reducing or preventing the digestive conversion of proteins into amino acids, which helps to deprive rapidly growing cancer cells of essential amino acids

In a mice model of colon cancer, carcinogenesis was suppressed, which may be due to cellular uptake of BBI that then facilitated inhibition of intracellular proteases involved in transformation from normal to malignant cells. In leukoplakia (human oral cancer), the anticancer effect of BBI is associated with inhibition of serine proteases that cleave the *neu* oncogene protein, which is a cell surface biomarker for human cancer. Inhibition of serine protease will cause accumulation of *neu* protein on the cell surface, which enhances immune recognition and destruction of the cancer cells by cytotoxic lymphocytes and natural killer cells. Oral administration of BBI has been shown to lead to detection in various organs within a 3-h period, which means that the protein is absorbed from the gastrointestinal tract. However, it is believed that BBI is more stable in the oral cavity because it is more resistant to hydrolysis in the saliva than in the other parts of the digestive tract. Therefore, BBI may exert longer therapeutic effects when present in the oral cavity when compared to the remaining sections of the digestive tract. Another plausible mechanism of anticancer action of BBI could be the fact that the protein or its complexes with proteases may act as free radical scavenger because of the high levels of cystine in BBI. By mopping up the potentially harmful free radicals, genetic materials (especially DNA) can be spared damage, which will help suppress carcinogenesis. The insoluble inhibitor-protease complex may act as an insoluble dietary fiber that physically adsorbs carcinogens as it passes along the digestive tract, which can prevent colon carcinogenesis.

Another anticarcinogenic compound identified from soybean seeds is lunasin, a 43-amino-acid protein. In animal models, lunasin has been shown to reduce the incidence of skin tumor and tumor yield with increased tumor latency period in comparison to untreated controls. Though a relatively big peptide, lunasin has been shown to be absorbed into the blood of rats and humans following ingestion of soybean proteins. One of the reasons that enhance bioavailability of lunasin is the presence of protease inhibitors in soybean products, which protects lunasin from structural breakdown by gastrointestinal enzymes and allows the protein to be absorbed into the blood circulatory system. The antioxidant properties of lunasin have been demonstrated using in vitro linoleic acid oxidation system where the protein and peptide fragments reduced the rate of lipid peroxidation. Lunasin and peptide fragment 1 (SKWQHQQDSCRKQLQGVNLTPC) also had free radical scavenging properties as determined using the ABTS^{+} radical cation assay. In lipopolysaccharide (LPS)-stimulated macrophages, the presence of lunasin or its peptide fragment 1 had no effect on cell viability but were able to attenuate reactive oxygen species (ROS) production. High levels of ROS produce oxidative stress, which could lead to changes in the cellular genomic information (DNA damage) and potentially induce carcinogenesis. In contrast, only lunasin was able to attenuate production of TNFα and IL-6 while the peptide fragments were inactive. Thus, the complete primary structure of lunasin seems to be required for the downregulation of cytokine production. Moreover, since immune potentiation might require binding to specific antibodies or cellular receptors, it is pos-

sible that specific folding pattern (secondary and tertiary structures) of the lunasin protein is required. Because fragmentation destroys the unique folding pattern, the ability to stimulate immune response was lost in the peptide fragments. Neither lunasin nor the peptide fragments had any effects on NO production by the LPS-stimulated macrophages. Lunasin can also suppress NF-κB-associated metabolic pathways, which together with its role as inhibitors of histone acetylation and retinoblastoma protein phosphorylation may provide a mechanism for chemopreventive properties.

5.3.4 Bone Health

Dietary soybean isoflavone has been associated with enhanced bone health, especially in menopausal women. Menopause is known to be associated with osteoporosis (reduced bone mineral density), which develops as a result of reduced plasma level of estrogen. Therefore, hormone replacement therapy with synthetic estrogen has been used as a therapeutic tool to reduce the pathological progression and intensity of osteoporosis. However, due to the potential negative side effects (e.g., breast cancer) of estrogen therapy in some women, attention has focused on the use of hormone mimetics such as soybean isoflavone for the treatment of osteoporosis. Initial epidemiological evidence showed that Southern and Eastern Asian women had significantly reduced incidence of osteoporosis-related bone fracture than Western women, a situation that was attributed to the high consumption of soy products in Asia. Analysis of various human intervention studies showed that dietary soybean isoflavones significantly reduced the urinary level of a peptide, deoxypyridinoline (Dpyr) marker for bone resorption. During bone remodeling, Dpyr is a peptide that is removed from the extracellular matrix, released into the blood circulatory system, and excreted in the urine. Therefore, high levels of Dpyr indicates high rate of bone matrix turnover, which can cause reduced mineral density. By reducing the amount of Dpyr released from the bone, soybean isoflavones act to reduce intensity of bone resorption and retard progression of osteoporosis. The effect of soybean isoflavones was stronger in postmenopausal than premenopausal women, suggesting an estrogen agonist effect under estrogen-depletion circumstances to provide bone beneficial effects. Bone-specific alkaline phosphatase (BAP) is a marker of bone formation that was significantly increased in the serum of women consuming soybean isoflavone. BAP is a non-collagenic protein that is produced by osteoblasts during bone matrix formation, and high serum levels are associated with increased bone mineral density. While the exact mechanism for the action of soybean isoflavones has not been deduced, it is possible that they stimulate estrogen receptors, especially those present in osteoblasts to enhance production of required extracellular matrix proteins. Bone health-promoting effect of isoflavones is probably mediated through increased production of insulin-like growth factor-1 (IGF-1). This is because IGF-1 is a recognized promoter of osteoblastic activity in human beings, and isoflavone-enriched diets have been shown to increase IGF-1 mRNA levels in the femur when compared to isoflavone-depleted diet. A high level of bone alkaline phosphatase is also associated with soybean isoflavone consumption, which is an indication of increased activity of the osteoblast. Overall, evidence point to the fact that soybean isoflavones inhibit bone resorption but stimulate bone formation, especially in postmenopausal women and could be used to delay the onset or reduce pathological intensity of osteoporosis. The beneficial effects of soybean isoflavones were associated with consumption of <90 mg/day during an intervention that lasted <12 weeks. However, because various levels of isoflavones have also been reported to have beneficial effects in clinical trials, there is a need for future trials that will provide better information on active dose, duration, and potential negative effects.

5.3.5 Menopause

The loss of estrogen during menopause leads to common symptoms such as hot flushes, night sweats, irritability, joint pains, irregular or heavy periods, and poor concentration. Phytoestrogens such as soy isoflavones may provide relief from these menopausal symptoms because of their ability to bind estrogen receptors and mimic estrogen activity. Therefore, hormone replacement therapy that uses soybean phytoestrogens may be used as substitute for synthetic estrogen in the management of menopausal symptoms. From epidemiological studies, it is known about 80% of European women experience hot flushes during menopause when compared to about 18% of Chinese women. It is believed that the lower incidence of hot flushes in Chinese women may be due to the high levels of soybean consumption. However, it is not yet known whether lifelong consumption of soybean products is necessary in order for women to experience the beneficial effects of soybean phytoestrogens. Several controlled human trials have shown lower incidences of hot flushes in groups that are fed soybean or isoflavone supplements. These studies also revealed that the beneficial effect of phytoestrogen consumption is highest in women with greatest severity of hot flushes. Moreover, daily diet supplementation with 25 g soybean protein containing 154 mg isoflavones led to reduction in subclinical atherosclerosis in healthy young women (<5 years postmenopausal) who were at low risk of developing cardiovascular diseases.

5.3.6 Nonalcoholic Fatty Liver Disease

As a result of metabolic imbalance associated with insulin resistance, there is increased level of plasma free fatty acids (FFAs), which accumulates in the liver to cause inflammation and fibrosis. The high circulating levels of FFAs induce excessive production of cytochrome P4502 E1 (CYP2E1) and increase lipid peroxidation-dependent and liver-damaging oxidative stress. Progression of the hepatic fatty acid accumulation can lead to a condition called "nonalcoholic steatohepatitis (NASH)" with or without cirrhosis. NASH is associated with insulin resistance and can be characterized by various metabolic disorders such as obesity, hypertension, diabetes, and hyperlipidemia. Using rat experiments that were fed a NASH-inducing diet, incorporation of soybean protein into the diet led to improved ratio of

HDL cholesterol/LDL cholesterol in addition to reduced plasma triglycerides and FFAs. The lower plasma FFAs was reflected in the reduced liver lipid accumulation, inflammatory cell aggregation, and fibrosis of the soybean protein-fed rats. The soybean protein-fed rats also had significantly lower plasma insulin, CYP2E1, and TNF-α, when compared to control rats that received no soybean protein. The active component in the soybean protein is unknown but could be peptides or amino acids that are released after digestion or could be the isoflavone components. A synergistic effect between all the protein components is also possible and could be focus of future research activities. Overall, soybean protein showed beneficial effects in reducing intensity of the pathological symptoms of NASH mostly through reduced oxidative stress, improved lipid metabolism, and increased sensitivity to insulin.

Bibliography

Aukema, H.M., and I. Housini. 2001. Dietary soy protein effects on disease and IGF-1 in male and female Han:SPRD-*cy* rats. *Kidney International* 59: 52–61.

Aukema, H.M., J. Gauthier, M. Roy, Y. Jia, H. Li, and R.E. Aluko. 2011. Distinctive effects of plant protein sources on renal disease progression and associated cardiac hypertrophy in experimental kidney disease. *Molecular Nutrition & Food Research* 55: 1044–1051.

Azadbakht, I., R. Shakerhosseini, S. Atabak, M. Jamshidian, Y. Mehrabi, and A. Esmaill-Zadeh. 2003. Beneficiary effect of dietary soy protein on lowering plasma levels of lipid and improving kidney function in type II diabetes with nephropathy. *European Journal of Clinical Nutrition* 57: 1292–1294.

Cahill, L.E., C.Y.-C. Peng, N. Bankovic-Calic, D. Sankaran, M.R. Ogborn, and H.M. Aukema. 2007. Dietary soya protein during pregnancy and lactation in rats with hereditary kidney disease attenuates disease progression in offspring. *The British Journal of Nutrition* 97: 77–84.

Castiglioni, S., C. Manzoni, A. D'Uva, R. Spiezie, E. Monteggia, G. Chiesa, C.R. Sirtori, and M.R. Lovati. 2003. Soy proteins reduce progression of a focal lesion and lipoprotein oxidability in rabbits fed a cholesterol-rich diet. *Atherosclerosis* 171: 163–170.

Dia, V.P., S. Torres, B.O. de Lumen, J.W. Erdman, and E.G. de Mejia. 2009. Presence of lunasin in plasma of men after soy protein consumption. *Journal of Agricultural and Food Chemistry* 57: 1260–1266.

Friedman, M., and D.L. Brandon. 2001. Nutritional and health benefits of soy proteins. *Journal of Agricultural and Food Chemistry* 49: 1069–1086.

Hernandez-Ledesma, B., C.-C. Hsieh, and B.O. de Lumen. 2009. Antioxidant and anti-inflammatory properties of cancer preventive peptide lunasin in RAW 264.7 macrophages. *Biochemical and Biophysical Research Communications* 390: 803–808.

Hodis, H.N., W.J. Mack, N. Kono, S.P. Azen, D. Shoupe, J. Hwang-Levine, D. Petitti, L. Whitfield-Maxwell, M. Yan, A.A. Franke, and R.H. Selzer. 2011. Isoflavone soy protein supplementation and atherosclerosis progression in healthy postmenopausal women. A randomized controlled trial. *Stroke* 42: 3168–3175.

Hsieh, C.-C., B. Hernandez-Ledesma, and B.O. de Lumen. 2010. Soybean peptide lunasin suppresses in vitro and in vivo 7,12-dimethylbenz-[a]anthracene-induced tumorigenesis. *Journal of Food Science* 75: H311–H316.

Hwang, S.-Y., C.G. Taylor, P. Zahradka, D. Bankovic-Calic, M.R. Ogborn, and H.M. Aukema. 2008. Dietary soy protein reduces early renal disease progression and alters prostanoid production in obese *fa/fa* Zucker rats. *The Journal of Nutritional Biochemistry* 19: 255–262.

Li, Y., D. Kong, B. Bao, A. Ahmad, and F.H. Sarkar. 2011. Induction of cancer cell death by isoflavone: the role of multiple signaling pathways. *Nutrients* 3: 877–896.

Ma, D.-F., L.-Q. Qin, P.-Y. Wang, and R. Katoh. 2008. Soy isoflavone intake inhibits bone resorption and stimulates bone formation in menopausal women: meta-analysis of randomized controlled trials. *European Journal of Clinical Nutrition* 62: 155–161.

Ogborn, M.R., E. Nitschmann, N. Bankovic-Calic, H. Weiler, and H.M. Aukema. 2010. Dietary soy protein benefit in experimental kidney disease is preserved after isoflavone depletion of diet. *Experimental Biology and Medicine* 235: 1315–1320.

Omoni, A.O., and R.E. Aluko. 2005. Soybean foods and their benefits: potential mechanisms of action. *Nutrition Reviews* 63: 272–283.

Peng, C.Y.-C., D. Sankaran, M.R. Ogborn, and H.M. Aukema. 2009. Dietary soy protein selectively reduces renal prostanoids and cyclooxygenases in polycystic kidney disease. *Experimental Biology and Medicine* 234: 737–743.

Siepmann, T., J. Roofeh, F.W. Kiefer, and D.G. Eldelson. 2011. Hypogonadism and erectile dysfunction associated with soy product consumption. *Nutrition* 27: 859–862.

Sirtori, C.R., C. Galli, J.W. Anderson, and A. Arnoldi. 2009. Nutritional and nutraceutical approaches to dyslipidemia and atherosclerosis prevention: focus on dietary proteins. *Atherosclerosis* 203: 8–17.

Swami, S., A.V. Krishnan, J. Moreno, R.B. Bhattacharyya, D.M. Peehl, and D. Feldman. 2007. Calcitriol and genistein actions to inhibit the prostaglandin pathway: potential combination therapy to treat prostate cancer. *The Journal of Nutrition* 137: 205S–210S.

Williamson, C. 2007. Health effects of soya- cause for concern? *Nutrition Bulletin* 32: 6–11.

Yang, H.-Y., Y.-H. Tzeng, C.-Y. Chai, A.-T. Hsieh, J.-R. Chen, L.-S. Chang, and S.-S. Yang. 2011. Soy protein retards the progression of non-alcoholic steatohepatitis via improvement of insulin resistance and steatosis. *Nutrition* 27: 943–948.

Fruits and Vegetables

6.1 Introduction

Generally, fruits and vegetables contain high levels of compounds and nutrients that may have beneficial effects on human health. For example, fruits and vegetables contain high levels of polyphenols that exhibit high antioxidant properties. Though a direct correlation between the antioxidant properties of fruit polyphenolic compounds and inhibition of cancer cell growth has not been conclusively proven, evidence suggests high levels of dietary fruits could reduce the incidence of cancer. Thus, other mechanisms apart from antioxidant effects may be responsible for observed antitumor effects of fruits and vegetables. For example, the polyphenolic compounds could inhibit certain enzymes that promote cell proliferation or promote activities of enzymes that induce cell death. However, it should be noted that the exact mechanism for the anticancer effects of fruits and vegetable polyphenols is yet to be fully determined. The effects could be direct action on cellular metabolism or indirect through stimulation of the immune system.

Epidemiological studies have also shown that increased consumption of soluble fiber in fruits has an inverse relationship with blood pressure in postmenopausal women. Specifically, fruit fiber consumption is directly related to plasma level of 16α-hydroxyestrone, a metabolite of the hormone 17β-estradiol, which is known to decrease with menopause. 16α-hydroxyestrone is a potent antioxidant that can increase production of prostacyclin (a vasodilator) at twice the rate of 17β-estradiol, endothelial nitric oxide synthase gene expression, nitric oxide production, and proliferation of vascular endothelial cell.

The polyphenolic compounds in fruits and vegetables have also been associated with blood pressure reducing effects, which is believed to occur through inhibition of angiotensin converting enzyme (ACE), a principal causative factor for hypertension. Specifically, the flavanol epicatechin and related oligomeric compounds (procyanidins) have ACE-inhibitory properties, in vitro. Among the oligomers, the tetramer is the most potent in vitro ACE inhibitor, though bioavailability and real effect in vivo remains to be demonstrated.

The following sections will illustrate potential beneficial effects on human health that has been reported for specific fruits and vegetables.

6.2 Ellagic Acid

One of the most common plant polyphenols called ellagic acid (Fig. 6.1) has been studied for potential health benefits, especially in relation to anticancer properties. Ellagic acid is an abundant polyphenol found in various fruits such as strawberry, raspberry, black currants, and grapes.

R.E. Aluko, *Functional Foods and Nutraceuticals*, Food Science Text Series,
DOI 10.1007/978-1-4614-3480-1_6, © Springer Science+Business Media, LLC 2012

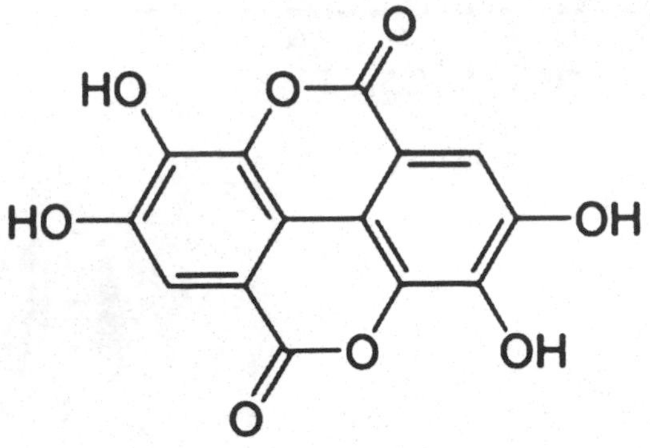

Fig. 6.1 Chemical structure of ellagic acid

6.2.1 Ellagic Acid and Cancer

One of the most common plant polyphenols called ellagic acid has been shown to exhibit antiproliferative and antioxidant properties during in vitro and in vivo tests. The antiproliferative properties of ellagic acid are related to ability to inhibit binding of some carcinogens to DNA, which can prevent induction of carcinogenesis. Similar to other antioxidant polyphenolic compounds, ellagic acid can reduce oxidative stress in cell culture models through radical-scavenging activities to produce chemoprotective effects. However, ellagic acid has been shown to have antagonistic effects against catechins, when present together. The potential chemopreventive properties of ellagic acid have generated interest in its use for modulating human health. Using a randomized controlled trial that involved 19 patients with carotid artery stenosis, it was found that pomegranate juice, which is high in ellagic acid, has potential benefits because of the observed reductions in blood pressure and carotid artery wall thickness. In prostate cancer patients undergoing chemotherapy, the use of ellagic acid supplementation reduced the rate of chemotherapy-associated loss of white blood cells, though there was no improvement in overall or progression-free survival of patients with prostate cancer. In male hamsters with 7,12-dimethylbenz[a] anthracene (DMBA)-induced hamster buccal pouch (HBP) tumors, addition of ellagic acid to the diet led to significant suppression of the carcinoma. In fact, at 0.4% inclusion of ellagic acid in the diet, the rate of tumor incidence was

zero, though there was a slight incidence of hyperplasia. The mechanism of action involved attenuation of the Wnt/β-catenin pathway by ellagic acid. The Wnt signaling pathway is a network of proteins that play important roles in pathogenesis by promoting embryogenesis and carcinogenesis. β-Catenin is a subunit of the cadherin protein complex and is an integral component of the Wnt signaling pathway. β-Catenin functions as part of the protein complex that constitute adherens junctions, which are required for the creation and maintenance of epithelial cell layers by regulating cell growth and adhesion between cells. β-Catenin is also important for transmitting the contact inhibition signal that stops cell division when the epithelial sheet formation is complete. Therefore, suppression of the Wnt signaling pathway can be used to suppress carcinogenesis and prevent progression and intensity of tumor pathology. Within the Wnt signaling pathway, expressions of Fz, Dvl-2, and GSK-3β were significantly high in DMBA-treated hamsters but were attenuated when ellagic acid was present in the diet. The nuclear expression of β-catenin was suppressed but that of cytosol was enhanced with ellagic acid supplementation. But nuclear translocation of the cytosolic β-catenin, which is required for activation of downstream target genes and signaling cascades, was attenuated because the required GSK-3β was present in low levels during ellagic acid supplementation. It should be noted that the Wnt signaling pathway is normally activated by binding of Wnt ligands to frizzled receptors (Fz), recruitment of cytoplasmic phosphoprotein disheveled (Dvl), and disruption of glycogen synthase kinase 3β (GSK-3β)-containing multiprotein complex. Thus, ellagic acid suppressed Wnt signaling pathway activation through attenuated expressions of Fz, Dvl-2, and GSK-3β. The DMBA-treated hamsters also showed increased expression of nuclear factor kappa B (NF-κB), which was suppressed when ellagic acid was present in the diet. The increased expression of NF-κB was due to the downregulated activity of inhibitory kappa B (IκB); IκB expression was increased by ellagic acid to prevent NF-κB activation. Ellagic acid supplementation increased the level of cellular apoptosis through upregulation of the expression levels of proapoptotic molecules like caspase 3 and 9. Moreover, attenuation of the Wnt and NF-κB pathways is known to enhance apoptotic cell death. Overall, it is evident that ellagic acid supplementation blocks or interrupts the functional cross talk between Wnt and NF-κB signaling pathways in HBP carcinomas, which leads to apoptotic cell death and prevention of carcinogenesis. Therefore, increased consumption of fruits and vegetables with high contents of ellagic acid may be beneficial for cancer prevention. Moreover, ellagic acid could be regarded as a potential compound for the formulation of anticancer foods and nutraceuticals.

6.2.2 Ellagic Acid and Cardiovascular Health

Ellagic acid has also been shown to have potential ameliorative effects on cardiac damage as evident by the protection of cardiomyocytes during drug-induced myocardial necrosis. Rats that did not receive ellagic acid showed neutrophil-infiltrated necrotic cardiac muscle fibers, whereas ellagic acid-pretreated rats had reduced myocardial necrosis with less edema and reduced neutrophil infiltration. In fact, hearts from the ellagic acid-pretreated rats had less edema and infarction with an almost normal myocardial architecture. The undesirable drop in blood pressure that accompanies myocardial infarction was significantly attenuated by ellagic acid treatment, with values that were very similar to normal rats. Oral administration of ellagic acid also significantly reduced oxidative stress markers such as C-reactive protein, plasma homocysteine, and lipid peroxides that are associated with myocardial infarction. Thus, ellagic acid has both antioxidant and anti-inflammatory properties. In contrast, cardiac levels of antioxidant enzymes such as superoxide dismutase and catalase as well as level of plasma antioxidant molecules such as α-tocopherol and vitamin C were significantly enhanced by ellagic acid treatment. In addition, the severity of heart damage as measured by leakage of cardiac enzymes such as troponin-I, creatine kinase, and lactate dehydrogenase was significantly attenuated by treatment

of the rats with ellagic acid prior to induction of myocardial infarction. Also important is the ability of ellagic acid to protect cardiomyocyte mitochondria during myocardial necrosis. The cardiomyocyte has a very high energy demand that is dependent on ATP from mitochondria oxidative phosphorylation; therefore, mitochondria damage could have significant negative effects on beat-by-beat contraction and relaxation of cardiac muscles. In addition to the antioxidant effects (increased antioxidant enzymes and reduced lipid peroxidation), ellagic acid pretreatment (7.5 and 15 mg/kg body weight) of rats led to increased activities of tricarboxylic acid enzymes (isocitrate dehydrogenase, succinate dehydrogenase, malate dehydrogenase, and α-keto glutarate dehydrogenase), which enhanced ATP formation when compared to the control rats that received no ellagic acid. The mitochondria of myocardial-infracted control rats contained high levels of cholesterol, free fatty acid, and triglycerides, all of which were significantly reduced as a result of pretreatment with ellagic acid prior to induction of myocardial infarction. High level of cholesterol in particular is undesirable in the mitochondria as it affects membrane fluidity and permeability to ions. During myocardial infarction, there is elevated level of mitochondria Ca^{2+}, which stimulates excessive production of reactive oxygen species and ultimately cell death. Rats that were provided oral administration of ellagic acid prior to induction of myocardial infarction had significantly reduced levels of Ca^{2+} in the cardiomyocyte mitochondria probably due to modulation of ion pumps. It is estimated that dietary consumption about 1.05 g of ellagic acid per day in the form of ellagic acid-rich fruits and vegetables (strawberries, raspberry, pomegranates, etc.) may be enough to produce these protective antioxidant and ATP-generating effects as seen during ellagic acid treatment of myocardial infarction.

6.3 Raspberries

The aqueous extract of red raspberries was obtained from a homogenate of the fruits that has been passed through a 0.2-μm filter. Antioxidant and anticancer effects of the filtered raspberry fruit extract was tested using stomach, colon, pancreatic, and breast cancer cell lines as well as other in vitro assays. All the cell lines showed susceptibility to berry extract-induced death, but the colon and stomach cancer cells were more susceptible than the breast cancer cells. Though all the cell lines were susceptible to vitamin C-induced cell death, the breast cancer cells were the most susceptible to this compound. It was evident that the antioxidant properties of vitamin C played a very strong role in the susceptibility of breast cancer and pancreatic cancer cells to berry extract-induced cell death. Though apoptotic death was not observed, cells that were treated with berry extract showed condensation of nucleus materials (chromatin) in addition to disorganization of the cytoskeletal structure. Thus, the berry extract had negative effects on cellular organization, which may have led to the observed cell death. It has been suggested that autophagy (catabolic degradation of cellular components) may play a role in the berry extract-induced death of the cancer cells.

6.4   Cherries

These are fruits of the *Prunus* genus within the Rosaceae family and typical examples include sweet cherry (*P. avium*) and tart cherry (*P. cerasus*). The fruits are rich in nutrients such as vitamin C, carotenoids, various polyphenolic compounds (especially anthocyanins), and fiber. Anthocyanins are glycosides of anthocyanidins and are compounds responsible for the red-purple color of sweet cherries. Some of the health benefits associated with cherry consumptions are described as follows.

6.4.1   Cardiovascular Effects

Most of the evidence for the health benefits of cherries has come from in vitro and animal studies. For example, an extract of tart cherry seed was found to improve regular heart beat and led to reduced cardiac damage in rat hearts that have been subjected to ischemic injury. Exposure of

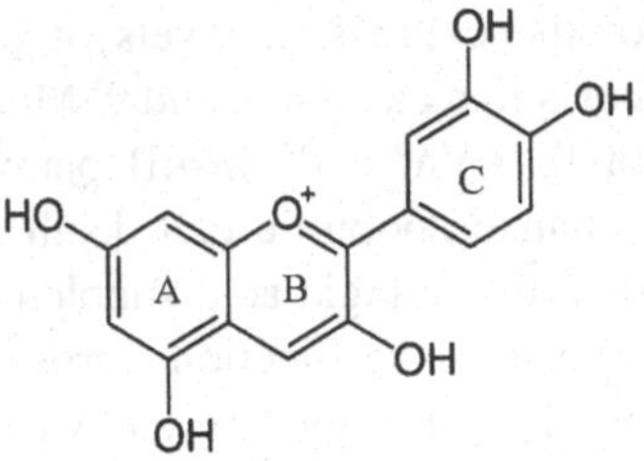

Fig. 6.2 Typical chemical structure of cyanidins showing the A, B, and C skeletal ring arrangement

bovine artery cells to cyanidin-3-glycoside (an anthocyanin) from cherry led to increase production of nitric oxide (NO), the vasodilatory agent. In addition to the upregulation of NO, there was decreased inflammation and reduced foam cell formation, which lowers the risk for development of atherosclerosis. Cyanidin-3-glycoside also reduced the level of cholesterol in macrophages and associated foam cells, which is an indication of ability of this cherry glycoside to lower the risk of cardiovascular disease.

6.4.2   Anti-inflammatory Effects

It is a well-known fact that low-grade inflammation can enhance the potential for development of various chronic diseases such as obesity, arthritis, kidney malfunction, cardiovascular disease, diabetes, and cancer. One of the main demonstrated effects of cherry components is the inhibition of cyclooxygenases (COX), the enzymes responsible for inflammatory response. Cyanidin and malvidin have the greatest anti-inflammatory effects among the cherry nutrients. Cyanidin (Fig. 6.2) is particularly very effective as an anti-inflammatory agent, which is probably because the structure contains a hydroxyl group on the B-ring. The extra hydroxyl group provides additional electrons that can be used to scavenge free radicals in addition to stabilizing the polyphenolic ring through increased ability to form resonance structures. In fact, cherry anthocyanins can inhibit COX-1 as much as 60% of the inhibitory level demonstrated by medications such as ibuprofen and naproxen while inhibition of COX-2 is actually higher than those exhibited by these medications. Cherry extracts (40 mg/kg) reduced the level serum tumor necrosis factor alpha (TNFα) and prostaglandin E2 (PGE2) in mice arthritis model, suggesting the potential use in treatment of this inflammatory condition. And in a human intervention trial that fed 280 g of sweet cherries daily for 4 weeks to healthy adults, there were significant reductions in the level of serum C-reactive protein, suggesting reduced inflammatory status when compared to baseline condition. What was more interesting is the increase in serum C-reactive protein when the intervention with cherries was terminated, which indicates that substances in the fruit were responsible for the observed anti-inflammatory effects.

6.4.3   Anticancer Effects

The anticancer effect of cherries is probably due to several nutrients such as fiber, ascorbic acid, carotenoids, and even the anthocyanins. While the levels of non-anthocyanin compounds are small and may be present in insufficient quantity within the fruit to produce anticancer effects, additive (synergistic) effects may contribute to increased potency against tumor development. However, the main anticancer compound in cherries is believed to be the anthocyanins and dark-red fruits will have higher levels than light-red fruits. Cherry diet, anthocyanins, and cyanidin have all been shown to reduce the number of cecal tumors but not colon tumors during mice feeding experiments, which suggest that the polyphenolic compounds may only be bioavailable in the cecum where it exerts antitumor properties. Antioxidant effects of cherry anthocyanins may also be a mechanism involved in the antitumor effects since cyanidin and the glycoside have shown protective effects on DNA cleavage when tested in cancer cell lines. Another mechanism proposed include evidence from cell culture, which showed that cherry anthocyanins induced apoptosis through arrest of the G2/M growth cycle. Cherry cyanidin also act as anticancer agent through increased radical-scavenging activity and inhibition of xanthine oxidase (enzyme that generates reactive oxygen species) activity.

Other mechanisms include ability of cherry nutrients to inhibit epidermal growth factor (EGF), and specifically, cyanidin can promote cellular differentiation, which reduces the risk for formation of malignant cells. Inhibition of EGF is particularly important since presence of this protein enhances cellular differentiation, proliferation, and ultimately survival. Therefore, by inhibiting EGF (probably through protein-polyphenol complex formation), cherry anthocyanins and anthocyanidins can enhance apoptosis and limit survival of tumor cells.

6.4.4   Antidiabetic Effects

The role of cherries as antidiabetic agents is believed to be due primarily to the antioxidant properties, high fiber content, and blood glucose-reducing properties exhibited by the fruit nutrients. This is because diabetes and associated symptoms is related to poor glucose utilization in addition to oxidative stress. Therefore, antioxidant compounds in cherries such as anthocyanins and quercetin may be able to reduce the risk of diabetes onset as well ameliorate pathological symptoms that occur during disease progression. In addition to antioxidant role, the cherry anthocyanins and anthocyanidins have been shown to reduce insulin resistance and glucose intolerance. In animal model of hyperglycemia, addition of cherry anthocyanins to the feed led to reductions in blood glucose. Insulin production can be enhanced in response to various glucose loads by pretreatment with cherry anthocyanins and anthocyanidins, which facilitates reduction in blood glucose levels. Moreover, sweet cherries has a lower glycemic index (22) and is a better fruit snack for diabetic patients when compared to high glycemic index fruits such as peach (42), plum (39), grapes (46), and apricot (57).

6.5   Grape Seed

Grape seeds contain several polyphenolic components, especially the proanthocyanidins or condensed tannins that are considered to possess health-promoting properties. In vitro work has showed that the ethanolic extract of grape seed can inhibit activity of some lipid-metabolizing enzymes such as pancreatic lipase (PL), lipoprotein lipase (LPL), and hormone-sensitive lipase (HSL). PL is responsible for lipid digestion, specifically breakdown of dietary triglycerides into free fatty acids and 2-monoacyl glyceride, all of which contribute to increase weight of adipose tissue and indirectly, obesity. Inhibition of PL by grape seed extract (GSE) could limit gastrointestinal breakdown of dietary triglycerides, reduce fatty acid absorption (triglycerides are not absorbable) and may contribute to reduced adipose tissue weight. LPL hydrolyzes lipoproteins to release triglycerides, which are then stored in adipocytes; therefore, LPL activity also contributes to obesity development. GSE also showed high level of LPL inhibition, which may contribute to reduced availability of triglycerides and reduced adipocyte storage. HSL hydrolyzes adipocyte fats and releases free fatty acids (FFA) into the blood circulatory system; high levels of plasma FFA is believed to induce insulin resistance and development of metabolic syndrome. Therefore, the ability of GSE to inhibit HSL activity may be used as a means of reducing plasma FFA content and reduce the risk of developing metabolic syndrome. Despite these potential health benefits, animal and human trials are required to validate bioactive properties of grape seed polyphenolic compounds as agents for preventing obesity and metabolic syndrome.

The immunomodulatory potential of grape seed proanthocyanidins (GSP) have also been studied and shown to potentiate antitumor effects in mice inoculated with Sarcoma 180 cells. The GSP had no direct cytotoxic effect on the tumor cells, but the antitumor effect was potentiated through stimulation of humoral and cellular immune response. This was evidenced by the increased thymus and spleen weight in addition to enhanced lymphocyte transformation, lysosomal enzyme activity, phagocytic capability of peritoneal macrophages, and production of tumor necrosis factor alpha (TNF-α) in GSP-treated mice. The increased weight of the thymus and spleen is reflective of activated immune response, which suggests that GSP could enhance an animal's immune system in order to attack and kill the tumor cells. High level of macrophage phagocytosis can promote tumor cell destruction while at the same time enhance production of TNF-α, a cytokine that can induce apoptotic tumor cell death.

6.6 Blueberries

The seeds of the *Vaccinium* plant are particularly very rich in phenolic compounds such as flavonoids, hydroxycinnamic acids, anthocyanins, and proanthocyanidins. The health benefits of blueberry consumption are associated mainly with the potent antioxidant properties of these polyphenolic constituents. For example, blueberry consumption is reported to protect against inflammation, improve cognitive functions, and attenuate disease progression of obesity/adiposity. Dietary supplementation of a high-fat mice diet with purified blueberry anthocyanins led to decreases in body and adipose tissue weights in addition to lower plasma levels of triglycerides, cholesterol, and leptin. There were also improvements in fasting blood glucose levels in addition to enhanced pancreatic β-cell function (insulin-producing cells found in the islets of Langerhans). The observed beneficial effects were not due to differences in food consumption as mice on blueberry and control diets had similar food intakes. But blueberry juice was not as effective as the purified anthocyanins except for the effect on reducing leptin levels. The lower efficacy of blueberry juice in reducing adiposity is probably due to attenuation of absorption of the bioactive anthocyanin molecules by other nutrients (sugars, polysaccharides, and lipids) present in the juice. In a separate experiment, freeze-dried whole blueberry powder did not reduce body or adipose tissue weight but improved insulin sensitivity and glucose homeostasis in mice. Mice that consumed the blueberry-supplemented diet had reduced adipocyte death, which was accompanied by downregulated gene expressions of TNF-α, IL-10, and CD11c (a surface marker for macrophages). This is significant because dead adipocytes cause an increase in the levels of proinflammatory cytokines such as MCP-1, TNF-α, and IL-10 due to the increased influx of macrophages. Thus, the blueberry extract was effective in reducing sensitivity and inflammation (attenuated cytokine levels plus decreased influx of macrophages to the adipose tissue) that is associated with obesity.

However, the antiobesity ability of the blueberry juice was substantially improved through biotransformation of the polyphenols with *Serratia vaccinii*, a bacteria present in the microflora of blueberry fruits. The biotransformed blueberry juice had increased phenolic content and antioxidant activities, but the negative effect on food intake may be responsible for the observed decreases in body weight gain, abdominal fat pads, and liver weight. Following chronic administration of the biotransformed juice to diabetic mice, there were reduced levels of plasma glucose and insulin levels, but adiponectin (an adipocytokine) level increased. The effect on adiponectin is significant because this cytokine have been shown to reverse insulin resistance in obese mice and can reduce muscle triglyceride levels through increased catabolism of free fatty acids. The biotransformed blueberry juice was also tested as an antioxidant agent for the protection of the nervous system and as potential therapeutic agent for the management of neurodegenerative diseases. The biotransformed juice significantly increased the activity of antioxidant enzymes (catalase and superoxide dismutase) and protected neurons against H_2O_2-induced cell death in a dose-dependent manner. The positive effects were associated with upregulation of mitogen-activated protein kinase (MAPK) family of enzymes such as p38 and c-Jun N-terminal kinase (JNK) as well as with the protection against H_2O_2-induced cell death through downregulation of extracellular signal-regulated kinase (ERK1/2) and MAPK/ERK kinase (MEK1/2) activities.

6.7 Strawberry

The health benefits of strawberry fruits have been associated with the very high contents of vitamin C, manganese, folate, and phenolic compounds, all of which contribute to improving oxidative status. The fruits also contain mostly the soluble form of dietary fiber, which can slow digestion and absorption of nutrients and enhance regulation of plasma blood sugar and insulin levels. The fiber in strawberry fruits can provide satiating and calorie-reduction effects, which contributes to reduced caloric intake. The seed oil is very high (~72%) in polyunsaturated fatty acids that can provide health benefits associated with consumption of omega-3 fatty acids. While the contributions of strawberry to healthy lipid consumption may be low due to the fact that the lipids are present only in the seeds, regular intake of the fruit could still make it a good source of essential fatty acids. However, the major bioactive components in strawberry are the polyphenolic compounds that include flavonoids (mainly anthocyanins), phenolic acids (hydroxybenzoic acid and hydroxycinnamic acid), lignans, hydrolysable tannins (ellagitannins and gallotannins), and condensed tannins (proanthocyanidins or procyanidins). The most important polyphenolic compounds are the anthocyanins (up to 800 mg/kg fresh fruit), which are responsible for most of the red color of strawberry fruits. Several (up to 25) strawberry fruit anthocyanin pigments have been reported, but the major one is pelargonidin-3-glucosides that includes pelargonidin-3-malonyl-glucoside, cyanidin-3-glucoside, and pelargonidin-3-rutinoside. Interest in strawberry fruits as functional foods has increased recently due to the reported bioactivities of the procyanidins, especially their role as antioxidant, antimicrobial, antiallergic, and antihypertensive agents. The main potential health benefits of strawberry anthocyanins are related to their antioxidant properties, especially ability to scavenge free radicals and prevent oxidative damage to blood lipids and DNA.

Strawberry glycosides can be absorbed intact or in the aglycone form after hydrolysis by β-glycosidases in the intestine. During absorption, the anthocyanins are conjugated with glucuronic acid such that the main metabolite present in the urine is pelargonidin-glucuronide as evident from human feeding studies. Pelargonidin-glucuronide was highest in the urine within the first 12 h of strawberry consumption and decreased thereafter. In humans that consumed strawberry fruits, serum lipoperoxidation was significantly decreased by

20% up to 22 h post-consumption. The ratio of 8-oxo-2'-deoxyguanosine (8-oxo-dG) to deoxyguanosine (dG) was decreased following strawberry consumption and returned to baseline values during the washout period when strawberry was withdrawn from the diet. 8-Oxo-dG is an oxidized dG product, and a high ratio reflects high level of DNA oxidation.

6.7.1 Anticancer Effects of Strawberry Fruits

Various cell culture and animal experiments have demonstrated the potential of strawberry fruits as anti-proliferation agents and in reducing the induction or disease progression of cancerous tumors. In a human feeding trial, formation of a carcinogen (N-nitrosodimethylamine) was shown to be reduced by 70% when strawberries were consumed immediately after intake of a diet rich in nitrates and amines. While the actual mechanism has not been elucidated, it is possible that the soluble fiber and polyphenolic compounds and carotenoids in strawberries bind and reduce availability of the carcinogen precursors. The presence of ellagic acid has been shown to be associated with the in vitro and in vivo chemopreventive abilities of strawberries as anticarcinogenic agents at the initiation and post-initiation stages of tumor development. Strawberry anthocyanins and tannins have also been shown to possess in vitro and in vivo antitumor properties when tested against various human cancer cell types. The mechanism of action of strawberry polyphenols is believed to be through their antioxidant properties, which inhibits mutagenesis and cancer initiation. These antioxidant properties include stimulation of antioxidant enzymes, decreased oxidative DNA damage, inhibition of carcinogen-induced DNA adduct formation, enhancement of DNA repair, and scavenging of reactive oxygen species. In addition to antioxidative effects, the strawberry polyphenols can also act as anticancer agents by promoting apoptosis and inhibiting angiogenesis, cell-cell communications, cell-cycle arrest, and inflammation. Modulation of other cell signaling pathways such

as cell proliferation and differentiation that is associated with cancer progression is also believed to be potential modes of action. For example, pretreatment of mouse epidermal cells with strawberry extracts had a dose-dependent suppression of activator protein-1 and nuclear factor kappa B activity that is normally induced by cancer-promoting agents like ultraviolet-B or O-tetradecanoylphorbol-13-acetate. Thus, the strawberry extract was able to inhibit proliferation and transformation of cancer cells and the associated signal kinase pathways. Other potential but yet to be elucidated mechanisms of cancer inhibition by strawberry includes induction of phase II-detoxifying enzymes, reduction in bioavailability of carcinogens, as well as inhibition of metalloproteinases and other enzymes involved in cancer metastasis.

6.7.2 Cardiovascular Effects of Strawberry Fruits

Initial data relating strawberry consumption to decreased risk of cardiovascular diseases came from the epidemiological study that involved overweight postmenopausal women. Increased level of strawberry consumption had an inverse relationship with cardiovascular disease (CVD) mortality after a 16-year follow-up. However, another study that involved a population with modest median strawberry consumption (1–3 servings per week) did not find any significant effect on the risk of CVD. But women who consume at least 2 strawberry servings in a week had significantly reduced risk of CVD when compared to those who do not consume strawberries. Therefore, the relationship between strawberry consumption and CVD risk reduction is related to the amount of the fruit that is consumed on a regular basis. A plausible mechanism of action of strawberry in reducing the risk of CVD is through the antioxidant properties of the fruit's constituents, especially polyphenolic compounds and vitamin C. In animal experiments, addition of strawberry fruits to the diet was shown to increase plasma antioxidant status with decreases in malondialdehyde formation. The strawberry group also had less DNA damage in the mononuclear blood cells, suggesting a strong antioxidative protective effect. In humans, long-term consumption of strawberries had significant effects of increasing LDL peroxidation lag time as well as increased erythrocyte resistance to oxidative damage. Strawberry polyphenolic compounds, especially anthocyanins, can act as antioxidants to protect lipid oxidation because they can be translocated into lipoprotein domains and cell membranes where interaction with lipid bilayers enhances protection of unsaturated fatty acids effects against oxidative agents. Within the vascular system, strawberry anthocyanins are incorporated into the endothelial cells and can be translocated into the cytosol. The presence of anthocyanins prevents oxidative damage to the endothelial cells and helps to preserve structure and function of the vascular system. In human intervention trials, daily consumption of dried strawberry extract or supplement led to reductions in plasma cholesterol and lipid (LDL) oxidation, thus supporting the ability of strawberry constituents to reduce in vivo oxidative stress. Another effect of strawberry is inhibition of α-glucosidase and α-amylase, which could enhance plasma glucose management by reducing starch digestion. Strawberry may also enhance blood pressure reduction because the fruit's extracts can inhibit angiotensin converting enzyme, a key enzyme involved in blood pressure elevation.

6.8 Blackberry

The high antioxidant plus anti-inflammatory capacities of blackberry polyphenolic extracts (mostly anthocyanins) has also been demonstrated, and there have also been various reports of the potential use as medicinal agents to reduce the risk of cancer and cardiovascular diseases. Using carbon tetrachloride (CCl_4)-induced oxidative stress in rats, oral administration of blackberry extract led to attenuated levels of hepatic lipid peroxidation with increased activities of antioxidant enzymes such as superoxide dismutase (SOD), glutathione reductase (GR), glutathione peroxidase (GPx), and catalase (CAT). In the liver, CCl_4 is converted to trichloromethyl free radical ($\cdot CCl_3$) by cytochrome P450 and then to trichloromethyl peroxide (Cl_3COO^-), both of which are highly reactive species. Hepatic damage is subsequently caused when $\cdot CCl_3$ or Cl_3COO^- attack and damage cellular components like nucleic acids, proteins, and unsaturated lipids. The attack on unsaturated lipids leads to formation of lipid peroxides, which breaks down into malondialdehyde (MDA), a common indicator of lipid peroxidation. Hepatic level of MDA was significantly increased by CCl_4 treatment but was reduced in rats that received the blackberry extract, which indicates ability of the polyphenols to scavenge the CCl_4-derived free radicals and prevent lipid peroxidation. During hepatic injury, as can be caused by high level of free radical species, there is increased cell membrane permeability that causes increased release of aspartate aminotransferase (AST) and alanine aminotransferase (ALT) into the blood. The CCl_4-induced free radical-mediated hepatic damage was attenuated by oral administration of the blackberry extract as indicated by reduced levels of serum AST and ALT. Molecular mechanism involved in the antioxidant effects of the blackberry extract involved upregulation of expression levels of nuclear factor E2-related factor 2 (Nrf2) and nuclear factor E2-dependent antioxidant enzymes such as SOD, GPx, and heme oxygenase-1 (HO-1). Nrf2 is an activating factor for the synthesis of antioxidant enzyme proteins because it positively regulates their mRNA expression. Therefore, blackberry polyphenolic extract-induced increase in the level of Nrf2 led to upregulated expression and protein synthesis of antioxidant enzymes, which enhanced hepatic antioxidant capacity.

Bibliography

Anitha, P., R.V. Priyadarsini, K. Kavitha, P. Thiyagarajan, and S. Nagini. 2011. Ellagic acid coordinately attenuates Wnt/β-catenin and NF-κB signaling pathways to induce intrinsic apoptosis in an animal model of oral oncogenesis. *European Journal of Nutrition*. doi:10.1007/s00394-011-0288-y.

Cho, B.O., H.W. Ryu, C.H. Jin, D.S. Choi, S.Y. Kang, D.S. Kim, M.-W. Byun, and I.Y. Jeong. 2011. Blackberry extract attenuates oxidative stress through up-regulation of Nrf2-dependent antioxidant enzymes in carbon tetrachloride-treated rats. *Journal of Agricultural and Food Chemistry* 59: 11442–11448.

Giampieri, F., S. Tulipani, J.M. Alvarez-Suarez, J.L. Quiles, B. Mezzetti, and M. Battino. 2012. The strawberry: composition, nutritional quality, and impact on human health. *Nutrition* 28: 9–19.

God, J., P.L. Tate, and L.L. Larcom. 2010. Red raspberries have antioxidant effects that play a minor role in the killing of stomach and colon cancer cells. *Nutrition Research* 30: 777–782.

Henning, S.M., N.P. Seeram, Y. Zhang, L. Li, K. Gao, R.-P. Lee, D.C. Wang, A. Zerlin, H. Karp, G. Thames, J. Kotlerman, Z. Li, and D. Heber. 2010. Strawberry consumption is associated with increased antioxidant capacity in serum. *Journal of Medicinal Food* 13: 116–122.

Mari Kannan, M., and S. Darlin Quine. 2011. Ellagic acid ameliorates isoproterenol induced oxidative stress: evidence from electrocardiological, biochemical and histological study. *European Journal of Pharmacology* 659: 45–52.

Mari Kannan, M., and S. Darlin Quine. 2012. Ellagic acid protects mitochondria from β-adrenergic agonist induced myocardial damage in rats: evidence from in vivo, in vitro and ultra structural study. *Food Research International* 45: 1–8.

McCune, L.M., C. Kubota, N.R. Stendell-Hollis, and C.A. Thomson. 2011. Cherries and health: a review. *Critical Reviews in Food Science and Nutrition* 51: 1–12.

Moreno, D.A., N. Ilic, A. Poulev, D.L. Brasaemle, S.K. Fried, and I. Raskin. 2003. Inhibitory effects of grape seed extract on lipases. *Nutrition* 19: 876–879.

Patel, S., L.C. Hawkley, J.T. Cacioppo, and C.M. Masi. 2011. Dietary fiber and serum 16α-hydroxyestrone, an estrogen metabolite associated with lower systolic blood pressure. *Nutrition* 27: 778–781.

Tong, H., X. Song, X. Sun, G. Sun, and F. Du. 2011. Immunomodulatory and antitumor activities of grape seed proanthocyanidins. *Journal of Agricultural and Food Chemistry* 59: 11543–11547.

Vuong, T., C. Matar, C. Ramassamy, and P.S. Haddad. 2010. Biotransformed blueberry juice protects neurons from hydrogen peroxide-induced oxidative stress and mitogen-activated protein kinase pathway alterations. *The British Journal of Nutrition* 104: 656–663.

7.1 Introduction

Apart from its nutritive values, milk contains various components that act as therapeutic agents to provide health benefits. These bioactive milk components include mainly the proteins, though some of the fats, sugars, and vitamins may also play important roles in human health. The main proteins are casein and whey, but there are several minor proteins and peptides such as lysozyme, transferrin, lactoferrin, hormones (insulin, somatostatin, adrenocorticotropin, prolactin, etc.), lactoperoxidase, enzymes, and immunoglobulins. For example, α-lactalbumin is rich in tryptophan and cysteine, two amino acids with important health attributes. Tryptophan and its metabolites regulate neurobehavioral effects such as appetite, pain perception, and sleep patterns. An α-lactalbumin-enriched product has been shown to improve sleep patterns, and a product containing this protein is currently marketed as a functional food for sleep aid. The presence of high levels of disulfide bridges in milk proteins is believed to stimulate the growth of *Bifidobacterium* species (bifidogenic effect), which are known to suppress growth of pathogenic bacteria in the colon. Lactoperoxidase is a protein with antibacterial activity and is found mostly in the colostrum, the milk produced in the first 3 days of lactation. In the presence of hydrogen peroxide, the enzyme catalyzes oxidation of thiocyanate to produce an intermediate compound that inhibits microbial growth. Milk lactoperoxidase can be used for cold sterilization of milk. Isolated lactoperoxidase can be added to commercial infant formula as a means of increasing the antimicrobial capacity of the food. Lysozyme is an antimicrobial enzyme present in colostrum and milk. The main bacterial cell wall constituents are susceptible to lysozyme-mediated hydrolysis, which leads to cell lysis and death. The enzyme is active against various Gram-positive and some Gram-negative bacteria and has a synergistic activity with lactoferrin against *E. coli*. Lactoferrin damages the outer cell membrane of Gram-negative bacteria, which makes the organism become susceptible to lysozyme. The antimicrobial effects may be exploited by fortifying foods with this enzyme as a means of reducing infections and improving human health. Milk products can also contain a glycomacropeptide (GMP), the 64 amino acid peptide released from κ-casein (169 amino acids) by rennet or chymosin during cheese making. Usually purified from the whey fraction, GMP is believed to play a role in satiety by stimulating increased production of cholecystokinin (CCK) in the intestine. This is because CCK is known to slow gastric emptying and can trigger intestinal stimuli without being absorbed. Apart from these minor protein components of milk, digestion of milk fat can release bioactive fatty acids. For example, digestion of milk fat in the stomach releases short- and medium-chain free fatty acids, which contribute to lowering of stomach pH. Low stomach pH facilitates digestion of proteins by enhancing activity of pepsin, the main digestive

R.E. Aluko, *Functional Foods and Nutraceuticals*, Food Science Text Series, DOI 10.1007/978-1-4614-3480-1_7, © Springer Science+Business Media, LLC 2012

enzyme in the stomach. Low stomach pH is also essential for promoting human health because it enhances the acid barrier against pathogenic microorganisms and creates a healthy stomach environment.

Other important protein components with health benefits are discussed as follows:

7.2 Whey Proteins and Anticarcinogenic Effects

This effect is due mainly to the high level of sulfur-containing amino acids, especially cysteine. Because whey proteins have higher levels of sulfur-containing amino acids, they are more effective than caseins as potential anticarcinogenic agents. While free cysteine is not readily absorbed, the disulfide form present in peptides has excellent bioavailability and has been shown to cause rapid increase in plasma concentrations of cysteine. During uptake into cells, the disulfide bond is cleaved, which allows entry of free cysteine molecules. Sulfur-containing amino acids are important because cysteine is the rate-limiting amino acid for the biosynthesis of glutathione (GSH), a natural antioxidant and a known anticarcinogen. GSH is responsible for scavenging of toxic free radicals and neutralization of highly reactive oxidative molecules that can damage DNA and cause abnormal cellular changes. By suppressing the damaging effects of free radicals and oxidants on the genes, GSH helps to prevent carcinogenesis and tumor growth. For example, in carbon tetrachloride (CCl_4)-treated rats, inclusion of whey proteins in the diet led to significant reductions in oxidative stress as evident from the low level of serum malondialdehyde and high total antioxidant capacity. CCl_4 produces hepatic toxicity that was attenuated by whey proteins as shown by the decreased levels of marker hepatic enzymes. CCl_4-treated rats showed markedly elevated hepatic fatty degeneration, necrosis, inflamed cells, and tissue apoptosis, which were all ameliorated when whey proteins were incorporated into the diet. GSH is also important for the immune response, and low level in HIV patients is predictive of poor survival. Evidence suggests that GSH enhances replication

and proper functioning of T cells that help fight and destroy foreign substances. Sulfur-containing amino acids are also important sources of sulfur which the liver can use to conjugate xenobiotics (foreign substances) and quench potentially toxic free radicals. Once conjugated with sulfur, the foreign substances (including drugs) become more soluble in the aqueous environment and less likely to enter the cells or tissues. The soluble sulfated substances are then easily excreted from the body through the urine; this mechanism prevents buildup of toxic foreign substances in the body.

7.3 Lactoferrin

Lactoferrin (Lf) is an iron-binding protein that has been shown to possess various properties that confer good health in humans. Because of its iron-binding capacity, Lf plays an important role in the regulation of iron homeostasis and can protect white blood cells against free iron-catalyzed oxidation reactions. Lf has activity against a broad microbial spectrum, including Gram-positive and Gram-negative bacteria, yeasts, and fungi. Lf also has antiviral activity, including activity against cytomegalovirus, influenza, and rotavirus. The mechanism of antimicrobial action of lactoferrin seems to involve the following:

- Due to the high affinity for iron, bacteria cells become iron deprived and stop growing. This is because enzymatic reactions and vital cellular processes that require iron for proper functioning become depressed or halted such that metabolic machinery becomes inoperable.
- Some microorganisms require iron as component of the cell membrane, or iron is required in reactions that maintain membrane integrity. Reduced availability of iron will lead to microbial membrane perturbation, and as a consequence the microbial cell loses its integrity and is killed.
- Stimulation of phagocytosis by macrophages and monocytes. By limiting availability of iron, microbial strength becomes weaker, and their destruction by phagocytosis is enhanced.

Several reports have demonstrated significant health outcomes following consumption of Lf as a therapeutic agent. Such improved health outcomes include increased suppression of colorectal adenomas, inhibition of hepatitis C virus replication in chronic conditions, enhanced activity of peripheral natural killer (NK) cell, and improved dermatological symptoms in athlete's foot (tinea pedis). From animal experiments, it has been shown that ingestion of Lf can reduce disease intensity during colitis, prevent cyclophosphamide-induced oocyte depletion, and inhibit hepatic inflammation that is associated with oxidative liver damage. The amelioration of colitis pathology was related to ability of Lf to correct underlying cytokine imbalance. Since Lf is a large protein, it is rapidly degraded during passage through the gastrointestinal tract; the whole protein is not likely to have direct interaction with target organs. Therefore, some of the bioactive properties of Lf probably come from low-molecular-weight peptide fragments that are small enough to be absorbed into the blood circulatory system. It is also possible that the whole protein or its digested fragments stimulate the immune system such that beneficial cytokines are released or undesirable cytokines become suppressed. For example, oral administration of Lf to mice led to increased production of interleukin-18 (IL-18) in the intestinal epithelium as well as type I interferon production in Peyer's patches; these products led to enhanced activity of NK cell. Enhanced activity of NK cell has been shown to inhibit mice tumor metastasis.

Lf has also been shown to stimulate production of IL-11 and bone morphogenetic protein 2 (BMP2) in addition to suppressing inflammation of the liver. IL-11 is a cytokine that is believed to suppress NF-κB activity, which results in decreased production of inflammatory cytokines. Therefore, IL-11 and BMP2 are both considered useful anti-inflammatory agents, which might be responsible for some of the health benefits associated with consumption of Lf. IL-11 may be used to ameliorate disease progression in patients suffering from hepatitis C-associated hepatic inflammation and advanced liver diseases. IL-11 also enhances cytoprotection by increasing production of heat shock protein in intestinal epithelial cells, which is important in maintaining gastrointestinal epithelial integrity and preventing gut-associated infections during chemotherapy in patients with blood malignancies. In patients with Crohn's disease, induction of IL-11 led to decreased inflammation and reduced disease intensity. Likewise, induction of BMP2 has been shown to enhance wound healing as well as suppress tumor growth by promoting apoptosis.

7.4 Colostrum, Immunoglobulins, and Growth Factors

Colostrum is also known as beestings or first milk and is the milk produced by the pregnant mammal during late pregnancy and in the first 2–3 weeks after delivery. Colostrum is a yellowish and thick milky fluid that is small in quantity but contains high levels of several essential nutrients and health-promoting factors. Colostrum also delivers essential compounds that help to strengthen the child's immune system; immune factors include immunoglobulins (Ig), cytokines, lactoferrin, lysozyme, and lactoperoxidase. In addition to the immune factors, colostrum also contains growth factors such as insulin-like growth factors (IGF), transforming growth factors (TGF), epithelial growth factors (EGF), platelet-derived growth factors (PDGF), and vascular endothelial growth factors (VEGF). These growth factors enhance normal body metabolism, ranging from synthesis and maturation of bones and muscles to expression of vital cellular genes. Furthermore, colostrum has a laxative effect that enhances the baby's ability to eliminate the first stool in order to clear dead red blood cells from the body and prevent jaundice. Colostrum plays an especially important role in the baby's gastrointestinal tract (GIT) as it promotes the buildup of beneficial bacteria in the intestines, which enhances digestion and absorption in addition to promoting growth and maturation of the GIT. The beneficial effects of colostrum are believed to enhance the survival, growth, and development of preterm babies. Previous studies have confirmed that colostrum intake promotes ability of neonates to establish passive immunity, support their metabolic and endocrine system, promote their GIT development and function, and generally improve their nutritional status and growth performance. For example, calves that were given colostrum in the first week after birth had improved metabolic and immunological status, especially with respect to lipid and protein metabolism, based on higher concentrations of plasma nutrients such as proteins, albumins, triglycerides, phospholipids, and cholesterol. The improved metabolism can be attributed to significant positive effects of colostrum on insulin, glucagons, and growth hormone. And in pregnant rats, oral administration of colostrum led to significant improvements in weight, length, and bone system development of the fetuses.

Ig molecules are usually glycoproteins produced by plasma cells, and they function as antibodies during immune response by binding with specific antigens. Five main classes of Ig have been recognized for their important roles in conferring immunity to humans: IgA, IgD, IgE, IgG, and IgM. Ig molecules are usually present in high concentrations in the colostrum but decline in concentration during normal lactation. Bovine colostrum contains 8–25% IgG, while human colostrum contains 2% IgG. The concentrations of IgA have been reported to have mean values of 3.5–4.5 mg/mL in human colostrum and an IgM value of 3.4 mg/mL, though the concentration varies very widely between individuals. The beneficial effects of colostrum vary from enhancing healthy maturation of the GIT to preventing pathogenic microbial infection. For example, in several developing countries, diarrhea is still an important factor that causes infant mortality with *Escherichia coli* (*E. coli*) as the principal causative factor. Oral administration of human or bovine colostrum has been shown to be effective in protecting against *E. coli* infection as well as treatment of infectious diarrhea. Several studies have demonstrated the ability of IgA from colostrum to substantially inhibit adhesion of *E. coli* to HEp-2 cells and limit the invasive capacity of the bacteria. Several components of colostrum, such as oligosaccharides, Ig molecules, and some non-immunoglobulin fractions, have also shown bacteria adhesion inhibitory activities. Previous works have shown that oral administration of colostrum might play an important role in elevating serum antibodies against *E. coli*. This is because studies have showed partial survival of the Ig molecules in the GIT, which suggests potential absorption into the blood circulatory system. Using an in vitro approach, it was shown that the IgA antibodies in colostrum were able to react with intimin, bfpA, EspA, and EspB, which are bacterial virulence-associated factors operating at different stages of enteropathogenic *E. coli* (EPEC) diarrhea. In conventional newborn calves that had restricted access to colostrum, it was shown that necrotoxigenic *E. coli* type 2 (NTEC2) strains were able to colonize the intestine, which led to long-lasting diarrhea and invasion of the blood stream. In immunodeficient patients infected with *Cryptosporidium parvum*, oral administration of bovine colostrum showed some promising results in limiting diarrhea, one of the main symptoms of the infection. Overall, colostrum has the following potential applications in diarrhea:

- Prevention of traveler's diarrhea
- Prevention of infections in day care centers and hospitals
- Secondary treatment in immunocompromised subjects such as AIDS patients and malnourished people as an agent to prevent microbial infection

Several studies have reported that colostrum may be an important functional food for the treatment of gastrointestinal disorders, including nonsteroidal anti-inflammatory drug (NSAID)-induced gut injury, inflammatory bowel disease, etc. This is because some of these research studies have shown that colostrum can ameliorate the severity of NSAID-induced gut injury, possibly as a result of the action of growth factors in the colostrum. In juvenile pig models, oral administration of colostrum resulted in normal weight gain, normalization of stool consistency, and enhanced morphologic adaptation after massive small bowel resection. And in a large proportion of patients that have distal colitis, oral administration of bovine colostrum led to rapid reduction in symptom scores and disease remission. In severe gut graft-versus-host disease (GVHD), clinical improvement after oral administration of colostrum has also been shown. Studies on the effect of various growth factors such as EGF, PDGF, TGF-β, or IGF-I in animals with colitis have shown potential use as therapeutic agents.

7.5   Milk Glycoproteins and Sugars

Bioactive milk sugars include oligosaccharides, mucin, gangliosides, and other N-acetylneuraminic acid-containing components. Though present in relatively small contents, they are important bioactive components because of their antimicrobial properties. For example, human milk glycoproteins can inhibit binding of enterohemorrhagic *E. coli* and prevent infection. The mannosylated glycoprotein form binds to the toxin produced by *E. coli* to inactivate the protein and prevent toxicity. The oligosaccharides provide antimicrobial activity against various organisms such as *Clostridia*, *E. coli*, and various pathogens by acting as prebiotics to enhance the population of colonic bifidobacteria. High *Bifidobacterium* populations produce a low pH colonic environment, which leads to suppression of growth of pathogens that are unable to tolerate high level of acidity. Apart from the prebiotic effect, the oligosaccharides enhance human immune response to rotavirus infection, which promotes rapid elimination of the pathogen. Oligosaccharides and especially mucins (high-molecular-weight glycoproteins) can also bind directly to bacteria cells such as *Campylobacter jejuni* and *Streptococcus pneumoniae* to inhibit colonization of the intestinal tract and prevent infection. Fucosylated oligosaccharides act as health promoters by binding directly to stable toxins produced by enterotoxic *E. coli*, which reduces toxic side effects while gangliosides bind to and neutralizes toxins produced by various enterotoxigenic microorganisms such as *Vibrio cholera*, *C. jejuni*, and *E. coli*.

7.6   Probiotics

Microbial fermentation is used to convert milk into yogurt that may contain billions of microorganisms. Probiotics are live microbial strains that impart health benefits through their influence on the gut microflora. Probiotics are presented to consumers in the form of fermented milk products (usually yogurt) or freeze-dried bacteria preparations sold as over-the-counter pills or tablets. Most common probiotic bacteria species include *Bifidobacterium*, *Lactobacillus*, *Streptococcus*,

and nonpathogenic *E. coli*. The role of probiotics is important because effective maintenance of the digestive system and immune system is influenced by a balance in the millions of bacteria that colonize the human gut. An imbalance that favors the growth and multiplication of the so-called healthy bacteria can enhance elimination of pathogenic bacteria to give a healthy gut. In contrast, poor diet, use of antibiotics, stress, foreign travel (where unfamiliar pathogenic strains can be ingested), and food poisoning can tilt the balance of colonic microflora in favor of the pathogenic strains, which can lead to disease. Therefore, consumption of foods that contain digestion-resistant probiotics will ensure that the bacteria arrive live in the colon, colonize it, and alter the microbial balance in favor of the healthy microbes. The growth and multiplication of probiotics usually benefit from the presence of certain carbohydrates in the colon. For example, the population of *Lactobacillus* and *Bifidobacterium* species increase when amylase-resistant starch is in the diet but enterobacteria population is decreased. *Bifidobacterium lactis* is the specie that shows the most rapid increase in population when the diet is rich in amylase-resistant starch. Based on this known specific bacteria growth-stimulating effects of certain carbohydrates, *synbiotic* products are prepared to contain both bacteria (probiotics) and saccharides (prebiotics). Synbiotics are designed to enhance survival of a particular group of beneficial bacteria because the specific substrate needed for fermentation and growth is provided, which provides competitive edge over non-probiotic microorganisms that are unable to ferment the prebiotic substrate. A good example of synbiotic preparation contains fructooligosaccharides and bifidobacteria. Following are key features of probiotics:

- Most often used bacteria cells are the bifidobacteria and lactobacilli.
- Can be consumed in fermented foods such as yogurt.
- Bifidobacteria and lactobacilli are normal inhabitants of the human colonic flora.
- Act by antagonizing pathogens and also by modulating the host immune defense mechanisms.
- A major goal of probiotic-based functional foods is the establishment of an upper bowel

(stomach and large intestine) microbial population that can act as defense against pathogens.

Criteria for isolating and defining probiotic bacteria:

- Must be of human origin
- Resistance to acidity and bile toxicity
- Adherence to human intestinal cells
- Colonization (even transient) of the human gut
- Antagonism against pathogenic bacteria
- Production of antimicrobial substances
- Immune modulation properties
- Clinically proven health effects (dose-response data)
- History of safe use in humans

7.6.1 Health Benefits of Probiotics

Probiotic bacteria will ferment dietary fiber to produce short-chain fatty acids (SCFAs), mainly acetates, propionate, and butyrates, which are compounds that increase blood flow to the colon and act as metabolic fuel for the intestinal cells. SCFAs are used for de novo glucose or lipid synthesis in addition to serving as energy source for the human host. Presence of SCFAs leads to lower colon pH that creates an unfavorable environment for the growth of harmful bacteria while promoting the growth of beneficial bacteria. The decrease in pH also facilitates absorption of minerals, especially calcium and magnesium, which enhances dietary nutrient utilization. Other mechanisms by which probiotics influence healthy gut development include:

- Immunity stimulation, especially through augmentation of natural killer cell activity and regulation of intestinal motility
- Elimination of pathogenic microorganisms through competition for limited nutrients or direct agglutination of pathogens
- Competitive blocking of adhesion or receptor sites on the cells of the mucosa surfaces
- Inhibition of epithelial invasion by pathogenic microbes through direct antagonistic effects
- Production of antimicrobial substances such as bacteriocins, organic acids, and hydrogen peroxide

- Release of gut-protective metabolites such as arginine, glutamine, short-chain fatty acids, and conjugated linoleic acids
- Binding and metabolism of toxic metabolites
- Mucus production
- Reduced gut luminal pH through stimulation of lactic acid-producing microorganisms

7.6.1.1 Effect on Gastroenteritis

Colonization of the gut by rotavirus is the most common cause of gastroenteritis in the world with associated symptoms such as diarrhea and vomiting. Anti-gastroenteritis effect of probiotics is evident in the reduced severity of rotavirus-induced diarrhea in children that consume foods containing *Lactobacillus casei*. Consumption of probiotics has been shown to boost immunoglobulin production as well as antibody response and other cellular immune responses against rotavirus, which would provide partial explanation for the reduced viral pathological symptoms. It should be noted that not all probiotics are effective, though one of the most effective strains has been identified as *Lactobacillus rhamnosus GG*. The use of probiotics to prevent traveler's diarrhea is not recommended so far because of the inconsistent results from various studies.

7.6.1.2 Coadministration with Antibiotics

Treatment of infections with antibiotics can lead to disruptions in the normal balance of microbial population in the gut and may lead to the reduction in population of healthy bacteria. Therefore, coadministration of probiotics with antibiotics or administration after antibiotic treatment have been investigated as a means of stabilizing the microbial population back to a health ratio and also to increase the immune response that can reduce the need for high doses of antibiotics. The diarrhea that is associated with use of some antibiotics may be ameliorated with administration of probiotics because the latter enhances formation of SCFAs from fermentation of dietary fiber. SCFAs stimulate water and sodium absorption in the colon, which enhances stool consistency and reduces severity of the antibiotic-induced diarrhea. Recent meta-analyses have confirmed that simultaneous or subsequent probiotic administrations can reduce the negative side effects associated with antibiotics treatment. However, these reported outcomes should be interpreted with caution until larger trials are conducted to confirm the earlier results.

7.6.1.3 Effects on Inflammatory Bowel Disease (IBD), Irritable Bowel Syndrome (IBS), and Other Gastrointestinal Disorders

IBD is a term that is used to describe diseases associated with the chronic and recurring inflammation of the digestive tract such as seen in Crohn's disease, ulcerative colitis, diverticulosis, and necrotizing enterocolitis. Probiotics may be used to treat IBD because of decreases in the growth of pathogenic bacteria, increased acidity (lower pH), prevention of tight junction invasion through enhanced barrier function, decreased expression of inflammatory markers, maintenance of remission, stimulation of immune responses, and lower drug consumption. Though very few clinical trials have been done, it is possible that the main beneficial effect of probiotics in IBD may be the prevention of damage to the intestinal mucosa following antibiotics administration or gastroenteritis. In particular, a preparation that contained mixed strains of lactobacilli and bifidobacteria is an effective agent in reducing symptoms of ulcerative colitis and pouchitis.

IBS is a digestive tract disease that is characterized by intermittent abdominal pains accompanied by alternating succession of diarrhea and obstipation including increased gas formation. IBS can be aggravated by gastroenteritis and antibiotic therapy, which disrupts the normal balance in colonic microbial population leading to reduced colonies of *Lactobacillus* and *Bifidobacterium*. In rats, diet supplementation with lactobacilli prevented visceral pain as well as stress- and antibiotic-induced visceral hypersensitivity. Results from recent clinical studies that involve probiotic use in IBS have been promising, though no definitive conclusion has been reached. For example, treatment with *Lactobacillus plantarum* 299v was able to relieve IBS symptoms while treatment with a probiotic

mixture of eight bacteria species led to improved abdominal bloating and reduced flatulence. Also, the treatment of IBS patients with bifidobacteria reduced abdominal discomfort and pain; the beneficial effects were associated with increased levels of anti-inflammatory cytokines, suggesting immune-modulating action as a possible mechanism. Thus, probiotics may serve as useful adjunct to conventional therapy for IBS patients through modification of gut flora that enables establishment of normal balance of gut microbial colonies.

Flatulence, constipation, diarrhea, and visceral pain are some of the gastrointestinal symptoms associated with stressful psychological events. Human experiments have shown that probiotics consumption can reduce nausea/vomiting and abdominal pain in addition to reduced flatulence and gas production. More importantly, there was no evidence that the probiotics treatment produced any negative side effects because the treatment was well tolerated. Consumption of probiotic yogurt has also been shown to reduce gastrointestinal discomfort (especially stomach pain) and the negative impact that associated symptoms have on patients using antiretroviral drugs.

7.6.1.4 Antiallergic effects

Current data suggests increasing incidence of allergy with about 20% of the population in western countries having one form of allergy or the other. Standard treatment usually involves avoiding the allergens, which provides relief for allergy sufferers but does not treat the disease. An alternative way of treating allergic reactions is to induce tolerance in patients through gradual exposure to the respective allergens. For ethical reasons, this may not be possible in severe cases, especially in children; therefore, the use of probiotics may offer a safer alternative. In infants, the immune system tends to be directed toward a T-helper (Th)2 phenotype, which is designed to prevent rejection of the fetus within the uterus. However, Th2 phenotype stimulates production of IgE by B cells, which increases the risk for allergic reactions through activation of mast cells. Microbial stimulation during early childhood development reverses the Th2 bias leading to stimulation of Th1 phenotype development and subsequent stimulation of the activity of Th3 cells. Consequently, IgA is produced by B cells, which contributes to allergen exclusion and reduced exposure of the immune system to antigens. Additionally, the Th1 phenotype produces cytokines that help to reduce inflammation and stimulate tolerance of common antigens.

Microbial stimulation of allergy resistance is supported by the fact that allergic children have an aberrant microbial population even prior to the onset of allergy. Allergic children have higher levels of clostridia and lower levels of bifidobacteria, which is accompanied by greater production of proinflammatory cytokines. The beneficial effects of probiotics in allergic subjects have been well demonstrated, but the exact mechanisms involved are not entirely known. From animal and cell culture studies, the following mechanisms have been observed or proposed:

- Modulation of intestinal microflora to favor growth and colony establishment of beneficial bacteria such as the *Bifidobacterium* and *Lactobacillus* species. This results in competitive exclusion of pathogenic bacteria that produce proinflammatory cytokines.
- Improved intestinal mucosa barrier function that leads to a reduction in leakage of antigens.
- Improved immune system through induction of anti-inflammatory cytokines or through increased production of secretory IgA, which acts to exclude antigens from the intestinal mucosa.
- Probiotic bacteria produce digestive enzymes that degrade dietary antigens to reduce the load of and exposure to antigens.

7.6.1.5 Anticancer Effects

In animal experiments, oral administration of *Lactobacillus casei* strain Shirota (LcS) had strong antitumor effects against transplantable experimental tumors. The same bacteria strain also inhibited tumor growth and migration in human malignant cells during clinical trials. Disruption of normal host immune parameters such as reduced secretion of interleukin-2 and low proportions of CD3$^+$, CD4$^+$, and CD8$^+$ T cells is involved in tumor promotion by some carcinogens. LcS works by preventing this carcinogen-mediated disruption of host immune parameters and maintenance of normal levels of various cytokines and T cells. LcS also works to prevent carcinogenesis by activating the natural killer (NK) cells, bone marrow-derived large granular lymphocytes that exhibit cytotoxicity toward a variety of tumors. Role of NK cells in cancer prevention is evident in human trials where medium to high activity is associated with reduced cancer risk, whereas low activity of the NK cells results in increased cancer risk. Proposed mechanisms of the anticancer effects of butyrates produced by probiotics include stimulation of cellular differentiation, inhibition of cellular proliferation, and enhanced apoptosis and histone protein acetylation (H3 and H4). Increased level of histone acetylation reduces the positive charge on histone proteins, which disrupts the ionic interaction with adjacent DNA backbone. Reduced histone-DNA interactions produce a less densely packed chromatin or euchromatin, which allows transcription factors to activate specific genes. Animal studies have shown a direct correlation between SCFA production and bacterial modulation of apoptosis, colonocyte proliferation, and differentiation. Evidence also shows that direct delivery of butyrate into the colon lumen reduced aberrant crypt formation. Probiotics also act as anticancer agents through increased conversion of linoleic acid to conjugated linoleic acid (CLA), which are known to possess anti-inflammatory properties, reduce cancer cell viability, and induce apoptosis. In particular, the 9-cis, 11-trans CLA has been found to inhibit development of carcinogen-induced aberrant crypt foci and number of polyps in mice. Probiotic-dependent metabolism of other nutrients such as isoflavones, lignans, flavonoids, and polyphenols may also be involved in cancer-preventive effects of beneficial microorganisms. Isoflavones are converted by probiotic bacteria in the colon to form equol, which has been shown to increase population of sulfate-reducing bacteria that helps lower the level of potentially carcinogenic compounds. Probiotics convert plant lignans into mammalian lignans (enterodiol and enterolactone) whose high concentrations have been associated with significant reductions in the risk for developing colorectal cancer. In cell culture studies, enterolactone was effective in inducing apoptosis, which inhibited growth of human colon cancer cells. Hops and hop-derived products contain prenylflavonoids such as xanthohumol, isoxanthohumol, and 8-prenylnaringenin (8-PN, a potent phytoestrogen). Growth of colon cancer cells were inhibited to a greater extent by xanthohumol than by 8-PN. Evidence also showed that 8-PN can inhibit proliferation of epidermal-growth-factor-induced MCF-7 breast cancer cells by modulating activity of phosphatidylinositol-3-OH kinase. With respect to polyphenols, probiotic bacteria are able to break down high-molecular-weight proanthocyanidins and oxidized polymeric equivalents (abundant in tea, chocolate, wine, and fruits), which normally are poorly absorbed in the small intestine. It has been postulated that the polyphenol metabolites have anticancer effects through decreased carcinogen-induced aberrant crypt formation, colonic proliferation, and oxidative DNA damage.

7.6.1.6 Antidiarrhea Effects

Increased liquidity or decreased consistency of stools that is normally associated with an increased frequency of stools and increased fecal weight is the major characteristic of diarrhea. The use of living bacteria to restore normal intestinal microflora has long been a therapeutic approach for treating diarrhea and associated symptoms. Yogurt was originally developed in Spain and sold as an affordable and easy to prepare remedy against diarrhea in children. The antidiarrhea effects of probiotic bacteria seen during viral or microbial infections could be due to their immunostimulatory properties or alleviation of symptoms and reduced intensity of acute infections, all of which have been confirmed in various clinical trials. The observed effects include reduced frequency of infections, shorter duration of episodes by up to 1.5 days, decreased shedding of rotaviruses, and increased production of rotavirus specific antibodies.

7.6.1.7 Effect on *Helicobacter pylori*

It is a well-known fact that long-term infection by *H. pylori* invariably causes chronic gastritis, which leads to peptic ulcer in addition to being a risk factor for development of gastric tumors. At present, the treatment of *H. pylori* infection involves expensive therapeutic interventions, which though are effective can also lead to development of unpleasant side effects and antibiotic resistance. Therefore, the use of food-based products such as probiotics may be a cost-effective way to manage symptoms associated with *H. pylori* infection. Various mechanisms, as summarized below have been observed or proposed for the observed beneficial effects of probiotics in *H. pylori* infections. Probiotics strengthen the defense attributes of the stomach by producing antimicrobial substances, competing with *H. pylori* for adhesion receptors, stimulating mucin production, and stabilizing the gut mucosa barrier:

- Antimicrobial compounds: bacteriocin and products of lactic acid fermentation, such as lactic and acetic acids, and hydrogen peroxide prevent the growth of *H. pylori*. Apart from lowering pH, lactate inhibits *H. pylori* urease, though this effect seems to be dependent on the strain of lactobacillus bacteria.
- Competition for adhesion to stomach wall: *H. pylori* infection and pathology is dependent on ability of the cells to adhere to the stomach epithelial cells. Inhibitory effects of probiotic bacteria on the adhesion capability of *H. pylori* are due to a combination of antimicrobial compounds and competition for adhesion sites. This competitive exclusion of *H. pylori* from the mucosa cell surface by probiotic bacteria usually involves nonspecific blockage of receptor sites. Moreover, regular consumption of probiotic foods could lead to a well-established stomach mucosa colony. Previous colonization of the mucosa cell surface will prevent or reduce *H. pylori* infection.
- Stomach mucosa barrier: during gastritis caused by *H. pylori* infection, there is reduced

secretion of mucus in the damaged epithelium, which enhances adhesion of the bacteria cells. Mucus secretion by the proliferating stomach epithelial cells is dependent on MUCI and MUC5A gene expressions, which are suppressed during *H. pylori* infection. Some lactobacillus strains can increase the expression of mucin genes, which enables restoration of the mucin-producing ability of gastric mucosa leading to reduced adhesion properties of *H. pylori*.

- Immune system: the body responds to *H. pylori* infection by releasing inflammatory mediators such as chemokines and cytokines. This response is initially manifested through a release of interleukin-8 (IL-8) that causes neutrophils and monocytes to migrate to the stomach mucosa. Subsequent responses involve monocytes and dendritic cells that produce tumor necrosis factor-α (TNF-α) along with other interleukins and interferons. These immune responses merely sustain the inflammation because it cannot remove the cause of infection, *H. pylori* cells. Probiotics act by modifying the host immune system through interactions with the epithelial cells to modulate production of anti-inflammatory cytokines such that gastric activity and inflammation are reduced. Probiotics also decrease the levels of specific IgG antibodies to *H. pylori* infection but stimulate local IgA responses that lead to strengthening of the mucosa barrier.

7.6.1.8 Effect on Mineral Absorption

Phytic acid in whole grain products binds to the cations and forms insoluble complexes that limit availability of various essential minerals present in the food. Availability of essential minerals such as copper, iron, and zinc is substantially depressed in humans and monogastric animals. By incorporating phytase-producing bacteria (e.g., *Mitsuokella jalaludinii*) into probiotic formulations, it is possible to increase mineral bioavailability in diets that contain high phytate levels. Direct effect on calcium absorption such as increased enterocyte uptake is a characteristic of the probiotic bacteria strain, *Lactobacillus*

salivarius (UCC 118). *Lactobacillus helveticus*-fermented milk enhances the formation of osteoblast bone, an effect that is believed to be due to release of bioactive mineral-binding peptides from milk protein by proteases produced by the bacteria during fermentation.

7.6.1.9 Antihypertensive Effects

Elevated blood pressure (hypertension) is a major independent risk factor for cardiovascular diseases. Likewise, high plasma levels of angiotensin converting enzyme (ACE) and blood lipids (LDL and triglycerides) can contribute to the pathogenesis of hypertension. Probiotics have been shown to modulate the intensity of these risk factors that are associated with hypertension. Rats that consumed *Lactobacillus casei*- and *Streptococcus thermophilus*-fermented milk had increased levels of HDL-cholesterol in addition to lower levels of serum triglycerides and atherogenic index when compared to control group. When tested in human subjects, consumption of the fermented milk produced decreases in systolic blood pressure. The fermented milk was shown to reduce ACE activity in vitro, which suggests a mode of action during blood pressure reduction. Another postulated mechanism is the fact that peptides liberated during bacteria fermentation of milk are capable of inhibiting cholesterol absorption in the jejunum by decreasing micellar solubilization of cholesterol. Also, the SCFAs produced during bacteria fermentation could reduce cholesterol absorption or the bacteria cells can bind and absorb cholesterol or reduce reabsorption of bile acids.

7.7 Role of Milk Fatty Acids in Cardiovascular Diseases

Several studies have shown the positive cardiovascular health effects of dairy product consumption in terms of improved blood lipid profile, decreased blood pressure, body mass index, and LDL/HDL ratio. The beneficial effects of dairy fat consumption has been be attributed in part to the presence of certain fatty acids such as tetradecanoic acid (C14:0), pentadecanoic acid (C15:0), and heptadecanoic acid (C17:0). It has been shown that a negative relationship exists between the content of two of the fatty acids (C14:0 and C15:0) in adipose tissue and the odds ratio for myocardial infarction (MI). And patients that consumed dairy products accumulated high levels of C14:0 and C15:0 in their adipose tissue and had lower BMIs. C15:0 and C17:0 have been used as valid markers for dairy fat intake, and their high levels in serum phospholipids are associated with reduced risk developing MI. The two fatty acid markers were also significantly negatively correlated with serum triglycerides, cholesterol, insulin, and leptin. Overall, elevated levels of serum C15:0 is believed to be protective against development of obesity in particular and metabolic syndrome in general. However, more extensive human intervention trials are required to clarify the importance and mechanism of action of these fatty acids as beneficial bioactive agents.

Bibliography

Biong, A.S., M.B. Veierod, J. Ringstad, D.S. Thelle, and J.I. Pedersen. 2006. Intake of milk fat, reflected in adipose tissue fatty acids and risk of myocardial infarction: a case-control study. *European Journal of Clinical Nutrition* 60: 236–244.

Collins, M.D., and G.R. Gibson. 1999. Probiotics, prebiotics, and synbiotics: approaches for modulating the microbial ecology of the gut. *The American Journal of Clinical Nutrition* 69(Suppl): 1052S–1057S.

Davis, C.D., and J.A. Milner. 2009. Gastrointestinal microflora, food components and colon cancer prevention. *The Journal of Nutritional Biochemistry* 20: 743–752.

de Vrese, M., and P.R. Marteau. 2007. Probiotics and prebiotics: effects on diarrhea. *The Journal of Nutrition* 137: 803S–811S.

Diop, L., S. Guillou, and H. Durand. 2008. Probiotic food supplement reduces stress-induced gastrointestinal symptoms in volunteers: a double-blind, placebo-controlled, randomized trial. *Nutrition Research* 28: 1–5.

Kuhara, T., K. Yamauchi, and K. Iwatsuki. 2012. Bovine lactoferrin induces interleukin-11 production in a hepatitis mouse model and human intestinal myofibroblasts. *European Journal of Nutrition* 51: 343–351.

Lesbros-Pantoflickova, D., I. Corthesy-Theulaz, and A.L. Blum. 2007. *Helicobacter pylori* and probiotics. *The Journal of Nutrition* 137: 812S–818S.

Li, H., and R.E. Aluko. 2006. Bovine colostrum as a bioactive product against human microbial infections and gastrointestinal disorders. *Current Topics in Nutraceutical Research* 4: 227–238.

Matsuzaki, T., A. Takagi, H. Ikemura, T. Matsuguchi, and T. Yokokura. 2007. Intestinal microflora: probiotics and autoimmunity. *The Journal of Nutrition* 137: 798S–802S.

Miles, L. 2007. Are probiotics beneficial for health? *Nutrition Bulletin* 32: 2–5.

Ouwehand, A.C. 2007. Antiallergic effects of probiotics. *The Journal of Nutrition* 137: 794S–797S.

Ramchandran, L., and N.P. Shah. 2011. Yogurt can beneficially affect blood contributors of cardiovascular health status in hypertensive rats. *Journal of Food Science* 76: H131–H136.

Scholz-Ahrens, K.E., P. Ade, B. Marten, P. Weber, W. Timm, Y. Asil, C.-C. Gluer, and J. Schrezenmeir. 2007. *The Journal of Nutrition* 137: 838S–846S.

Séverin, S., and X. Wenshui. 2005. Milk biologically active components as nutraceuticals: review. *Critical Reviews in Food Science and Nutrition* 45: 645–656.

Fish

8.1 Bioactive Components

Epidemiological studies have shown that the incidence of coronary heart diseases is low among populations that consume large quantities of fish or fish products. This beneficial effect is believed to be due to low incidence of ventricular fibrillation, which is commonly found during myocardial infarction. Fish oil consumption contributes to reductions in plasma levels of triacylglycerides, especially in patients suffering from hypertriglyceridemia. Dietary supplementation of hypercholesterolemic animals with fish oil reduces the incidence of atherosclerosis and stiffening of arterial walls. In a rat model of myocardial dysfunction, dietary fish oil increased the level of long-chain and n-3 polyunsaturated fatty acids (PUFA) in cardiac cells when compared to the control. Consumption of fish oil is also able to reduce the occurrence of platelet aggregation in the blood circulatory system. Some of the pathological changes associated with ischemic-reperfusion such as decreased force of cardiac contraction, increase in coronary perfusion pressure, appearance of ventricular arrhythmias, and release of creatine kinase and thromboxane B2 in the coronary effluent were all attenuated by fish oil supplementation. Therefore, dietary fish oil has a direct effect on the heart independent of blood and plasma components. The beneficial effects of fish oil in reducing postoperative body temperature mortality have also been demonstrated in clinical studies. However, apart from the oil, it

should be noted that other nutrients in fish such as proteins, minerals (copper, calcium, selenium, zinc, and magnesium), and vitamin B_1 may also have beneficial effects on human health. Overall, it is recommended that consumers include regular fish consumption (at least two portions per week) in their diets as a means of benefiting from the fish bioactive nutrients and for maintaining good health status. The use of dietary antioxidants is recommended during chronic use of fish oil due to the high propensity of the unsaturated fatty acids to form highly oxidized products. This is because the increased oxidative stress that is associated with fish oil consumption depletes endogenous tocopherol (an antioxidant), decreases cellular function, causes organ dysfunction, and leads to accelerated aging and shorter life span in senescence-accelerated mice.

8.2 Role of Fish Components in Specific Disease Conditions

8.2.1 Cardiovascular Diseases

High levels of triacylglycerides are directly correlated with high levels of blood clotting factors such as prothrombin, factor V, factor VII, and factor X in overweight and diabetic patients. This type of association increases the risk of blood clot formation and severe impairment of the normal functioning of the cardiovascular system. Consumption of fish oil reduces blood levels of

R.E. Aluko, *Functional Foods and Nutraceuticals*, Food Science Text Series,
DOI 10.1007/978-1-4614-3480-1_8, © Springer Science+Business Media, LLC 2012

triacylglycerides and decreases the associated risk of blood coagulation within the circulatory system. Proposed mechanisms for the beneficial action of fish oil in regulating blood level of tria-cylglycerides include:

- n-3 PUFA-dependent downregulation of gene expression by the liver
- Inhibition of production of VLDL apo B or stimulation of apo B breakdown
- Improved clearance of chylomicrons, the major transport vehicle for triacylglycerides
- Increased conversion of VLDL into LDL in peripheral tissues

Epidemiological studies have demonstrated a dose-related effect between regular dietary fish intake and a reduction in the incidence of coronary heart disease. For example, in men suffering from cardiac infarction, a 29% reduction in all-cause mortality within a 2-year period was observed when two fish meals (300 g) per week were introduced into their diet. This association was stronger in diabetic women where there was a 60% lowered risk of CHD in the group that had the highest fish intake when compared with groups with little or no fish in the diet. Even in patients that had already developed CHD, dietary intake of as little as 150 g of fish led to lower-risk of serious health complications. The main bioactive component responsible for the cardiovascular health benefits of fish seems to be the omega-3 PUFAs. This is because supplementation of patients' diet with omega-3 PUFAs was found to have significantly reduced the risk of cardiovascular deaths, sudden cardiac death, all-cause mortality, and nonfatal cardiovascular events.

8.2.2 Brain Function

The beneficial effects of fish consumption on the brain concern three main areas, which are depression, early human development, and maintenance of cognitive functions later in life. These beneficial effects of fish consumption are believed to be mediated through the omega-3 fatty acids that are present in the oil as well as through several intrinsic antioxidant compounds that naturally protect the PUFAs against oxidative damage. In patients

that suffer from depression, intervention with dietary fish has been found to lower symptoms. Epidemiological studies also showed lower prevalence of depression in populations with high annual fish consumption, while men that consumed low amounts of fish were more susceptible to developing depression. Similarly, high intakes of EPA and DHA have been shown to be associated with lower risk of depression.

The beneficial effects of fish consumption on fetal development and early life are related to the fact that large amounts of DHA are incorporated into the brain and retina. Due to the structural kinks introduced by the cis-double bonds, tight linear packing of DHA is impossible, and therefore, large amounts of this fatty acid lead to improved fluidity of cell membranes of the brain and retina. The DHA-induced fluid cellular structure is believed to enhance optimal functionality of the organs through increased ability of signaling agents (ions, eicosanoids, sugars, etc.) to translocate in and out of the cell as required. In animal experiments conducted with dietary restrictions of DHA, preterm infants develop poor cognitive and retina functions. However, when DHA-enriched diets are used, there is improved functioning of the brain and retina. When fish oil-supplemented diet is used during pregnancy, the infants developed better mental processing and problem-solving skills as well as improved hand/eye coordination. Using epidemiological studies, it was shown that mothers who consume at least one serving of fish four times in a week during pregnancy gave birth to babies that had higher developmental scores at 18 months when compared to infants from mothers that ate no fish. DHA has been shown to be present in breast milk of lactating mothers who ate fish on a regular basis and higher levels of milk. DHA content in breast milk is correlated with increased fish consumption. This is important because visual acuity in infants has been shown to have a positive correlation with the amount of DHA in breast milk.

In terms of maintenance of cognitive functions in later life, most of the interest has to do with ability of fish oil to mitigate cognitive decline, dementia, and Alzheimer's disease. Most of the data on the effects of fish oil on cognitive functions in later life has come from epidemiological studies. For example, Dutch adults that consumed more than 20 g of fish per day had reduced risk of developing cognitive decline, dementia, and Alzheimer's disease. Similarly, a 4-year follow-up study of elderly US people who have normal brain functions showed that there was a 60% reduction in the risk of Alzheimer's disease among the population that consumed oil-rich fish at least once per week.

8.2.3 Cancer

While there is limited data regarding the positive effects of fish consumption on cancer development and growth, evidence suggests that fish components may provide some protective effects against the pathogenesis of tumors. In particular, there is substantial scientific evidence that fish oil-containing diets rich in omega-2 PUFAs such as DHA and EPA can protect against tumorigenesis. For example, it has been shown that EPA in the diet led to reductions in rectal polyp numbers and size in patients that have familial adenomatous polyposis (FAP). In these patients, FAP was suppressed to a degree that is similar to the action of the drug, celecoxib (COX-2 inhibitor). The anticancer effects of fish oil can be greatly enhanced when diet also includes fermentable fiber. This is because butyrate produced from fiber fermentation has a synergistic effect with fish oil PUFAs, especially DHA (EPA was less effective) in maximizing the apoptosis of cancerous cells. In fact, DHA alone has minimal apoptotic effect, but coadministration with butyrate increased apoptosis by 43%. The action of DHA is believed to involve significant Ca^{2+} accumulation in the mitochondria, and upregulation of targeted apoptosis of DNA adducts during tumor initiation coupled with spontaneous apoptosis during tumor promotion. The mechanism responsible for the increased anticancer (apoptosis) activity of DHA in the presence of butyrate is through induction of p53-independent, oxidation-sensitive, mitochondrial Ca^{2+}-dependent pathway. The high concentration of DHA will increase mitochondria oxidative activity to generate high levels of reactive oxygen species and lipoperoxides, which stimulates Ca^{2+}

influx into the mitochondria. It has been shown that butyrate potentiates DHA-induced release of Ca^{2+} from intracellular stores and the resultant mitochondria loading. The high mitochondria Ca^{2+} level interacts with cyclophilin D, which triggers opening of the permeability pore to allow entry of more ions. The disturbance in the mitochondria Ca^{2+} homeostasis eventually induces series of reactions that lead to apoptotic cell death.

It is believed that fish oils and vitamin D may enhance ability of colon epithelial cells to undergo proper differentiation, which lowers the risk of mutation. In some populations, high dietary fish levels have been found to be associated with reduced risk incidence of colon cancer. However, the evidence seems to point at a preventive effect rather than a curative effect. This is because in populations that already have polyps in their colon and therefore are at high risk of developing colon cancer, dietary intervention with oil-rich fish did not reduce the levels of cancer risk markers.

8.2.4 Immune System

One of the main effects of fish oil components on the immune system is that high plasma levels of anti-inflammatory EPA and DHA (resulting from high dietary fish intake) can lead to increased incorporation into immune cells. As the immune cells accumulate EPA and DHA, the proinflammatory fatty acids such as omega-6 PUFAs and arachidonic acid become displaced, which reduces the risk for inflammation. For example, consumption of 150 g of salmon three times a week led to reductions in proinflammatory markers such as C-reactive protein, interleukin-6, and prostaglandins, which indicates improvements to the immune function. Mothers who consumed high levels of oil-rich fish during pregnancy had lower risk of food intolerance and produced offsprings with lower risk of wheeze. In adults, high dietary fish is associated with lower risk of asthma or allergy development, while lack of fish during childhood led to increased risk of asthma development. Using IgE levels and lung functions, it was found that high level of dietary fish and DHA was associated with a reduced rate of allergic sensitization.

8.2.5 Diabetes

Data from several animal experiments have shown that dietary fat-free fish proteins could be used to improve glucose management and reduce insulin resistance. For example, addition of cod protein to a high-fat, high-sucrose diet had protective effects against development of obesity-linked insulin resistance and glucose tolerance. But the oral glucose tolerance test did not detect any beneficial effect of cod protein when compared to casein. However, in terms of insulin sensitivity, the amount of glucose required to cause hyperglycemia was significantly higher for the cod-fed rats than the casein-fed rats. Thus, the cod protein enhanced insulin-dependent glucose uptake, which suggests improved insulin sensitivity when compared to casein. The increased insulin sensitivity was associated with reduced level of fat accumulation and weight gain. When the cod protein was also fed to human subjects, similar beneficial effects on insulin and glucose management were observed. In insulin-resistant human subjects, consumption of cod protein improved insulin sensitivity and was associated with a reduction in plasma C-reactive protein (CRP), an inflammatory marker that is elevated in this disease condition.

8.2.6 Obesity

The use of salmon as the protein source in a high-fat, high-sucrose rat diet led to significant weight gain reduction in comparison to casein. Since the two rat groups had similar caloric intake, the weight reduction suggests that the salmon protein enhanced energy expenditure when compared to casein. Moreover, the salmon diet prevented dietary fat-mediated increase in the weight of epididymal white adipose tissue (WAT) when compared with casein. Increase weight of WAT is a known risk factor for development of obesity-linked insulin resistance, which is linked to increased water retention. Therefore, apart from the increased expenditure, reduction in weight gain associated with fish diet could be because of the increased insulin sensitivity, which reduces

water retention in the body. The mechanism responsible for the increased energy expenditure could involve the activity of salmon calcitonin (sCT), a bioactive protein that upregulates energy expenditure. Rats that consumed the salmon diet had significantly higher plasma levels of sCT, which supports the hypothesis that weight reduction was associated with increased energy expenditure when compared to the casein diet. In comparison to casein, various fish proteins have been found to reduce the level of proinflammatory cytokines (TNF-α and IL-6) in WAT, which is probably due to reduced fat accumulation. However, only salmon protein was effective in reducing macrophage-associated inflammation. While data is still emerging, consumption of fish has been shown in some human intervention trials to result in significant weight loss when compared to nonfish diets of overweight adults. Other works have shown little or no benefits to weight loss when the dietary fish is used. However, it should be noted that high protein diets may have satiating effects, which could be responsible for some of the weight loss associated with fish consumption. This is because the satiating effects will reduce food and total caloric intake while the proteins become incorporated into lean muscles when compared to dietary fat-induced adipose tissue buildup.

8.2.7 Kidney Disease

Fish oil has been tested as a potential therapeutic intervention agent against chronic kidney disease (CKD). Use of fish oil has been mostly as an anti-inflammatory agent to reduce the high degree of inflammation associated with CKD as a result of high levels of TNF-α and other inflammatory markers. When peripheral blood mononuclear cells from CKD patients were treated with fish oil, there was decreased production of tissue factor (TF) procoagulant activity, which indicates reduced inflammatory response. TF is known to trigger in vivo blood coagulation with high levels associated with increased risk of developing blood clots and atherosclerosis. Various in vivo and cell culture tests have also shown that fish oil reduced IL-6 and monocyte chemoattractant protein-1

(MCP-1) secretions. MCP-1 is a proinflammatory cytokine that plays a role in monocyte recruitment of early atherosclerotic lesions, and level of this compound is increased as the CKD progresses. High level of MCP-1 leads to increased intensity of CKD symptoms, especially plaque formation within vascular walls. Therefore, the ability of fish oil to reduce MCP-1 levels is beneficial towards reducing the severity and progression of cardiovascular disease in CKD patients. Fish oil also reduced levels of CRP, a well-known predictor of cardiovascular complications and morbidity in CKD patients. The beneficial effects of fish oil in reducing the risk of major cardiovascular events such as blood clot and plaque formation are particularly very important since many CKD patients actually die from the cardiovascular complications associated with the disease.

In rats subjected to obstructive renal injury, inclusion of fish oil in the diet reduced inflammation (attenuated macrophage infiltration), renal collagen deposition, expression of transforming growth factor-β1 (TGF-β1), and apoptosis. The fish oil also attenuated production of other proinflammatory cytokines, including MCP-1, leukotriene B$_4$, and IL-1β. The reduced expression of cytokines in the fish oil diets was probably responsible for the decreased macrophage infiltration and reduced inflammatory status. Less macrophage infiltration (decreased inflammatory response and oxidative stress) had favorable effects in reducing rate of apoptosis. It is well known that macrophages produce free radicals and oxidative species that can induce DNA damage and cell death. Rats that were fed with fish oil-containing diets had increased expressions of heme oxygenase 1 (HO-1), an inducible enzyme that catalyzes heme degradation. HO-1 is an isoform that is produced in response to stress conditions such as oxidative stress, hypoxia, and cytokines. Following urethral obstruction, HO-1 enzyme plays an important role in renal physiology as an antiapoptotic agent. Fish oil also reduced renal fibrosis probably as a result of decreased expression of TGF-β1 and TNFα, both of which are involved in the formation of fibrosis and collagen. Overall, fish oil consumption may be particularly beneficial in reducing disease associated with urinary tract obstruction, especially in high-risk groups like the elderly population.

8.2.8 Digestive Tract System

Consumption of fish oil provides beneficial effects in the management and treatment of various digestive tract diseases such as inflammatory bowel conditions, ulcerative and Crohn's diseases. The mechanism involves modifications of the fatty acid composition of cell membranes, especially by the n-3 polyunsaturated fatty acids (n-3 PUFA). In the colon and ileum, eicosapentaenoic, docosapentaenoic, and docosahexaenoic acids become increasingly incorporated into cell membrane phospholipids primarily at the expense of the n-6 PUFA (linoleic and arachidonic acids). Incorporation of n-3 PUFA molecules into cell membranes leads to positive functional changes in membrane receptors, ion channels, and second messenger systems. There is also increased contraction of the ileum but not the colon when fish is part of the diet. Increased levels of n-3 PUFA in the cell membranes lead to structural effects that arise from physicochemical changes in the membrane environment as well as modulation of gene expression. These physicochemical and functional changes lead to modification of endogenous mediator profiles such as eicosanoids in addition to influencing physiological responses to exogenous agonists. Consumption of fish oil also leads to decrease in cecal digesta pH and increase in the level of butyrate, which contribute to better bowel health.

Bibliography

Kolar, S.S.N., R. Barhoumi, J.R. Lupton, and R.S. Chapkin. 2007. Docosahexaenoic acid and butyrate synergistically induce colonocyte apoptosis by enhancing mitochondrial Ca2+ accumulation. *Cancer Research* 67: 5561–5568.

Kolar, S.S.N., R. Barhoumi, C.K. Jones, J. Wesley, J.R. Lupton, Y.-Y. Fan, and R.S. Chapkin. 2011. Interactive effects of fatty acid and butyrate-induced mitochondrial Ca2+ loading and apoptosis in colonocytes. *Cancer* 117: 5294–5303.

Patten, G.S., M.Y. Abeywardena, E.J. McMurchie, and A. Jahangirl. 2002. Dietary fish oil increases acetyl-choline- and eicosanoid-induced contractility of isolated rat ileum. *The Journal of Nutrition* 132: 2506–2513.

Peake, J.M., G.C. Gobe, R.G. Fassett, and J.S. Coombes. 2011. The effects of dietary fish oil on inflammation, fibrosis and oxidative stress associated with obstructive renal injury in rats. *Molecular Nutrition & Food Research* 55: 400–410.

Pilon, G., J. Ruzzin, L.-E. Rioux, C. Lavigne, P.J. White, L. Froyland, H. Jacques, P. Bryl, L. Beaulieu, and A. Marette. 2011. Differential effects of various fish proteins in altering body weight, adiposity, inflammatory status, and insulin sensitivity in high fat-fat-fed rats. *Metabolism* 60: 1122–1130.

Ruxtin, C.H.S. 2011. The benefits of fish consumption. *Nutrition Bulletin* 36: 6–19.

Shing, C.M., M.J. Adams, R.G. Fassert, and J.S. Coombes. 2011. Nutritional compounds influence tissue factor expression and inflammation of chronic kidney disease patients in vitro. *Nutrition* 27: 967–972.

Tsuduki, T., T. Honma, K. Nakagawa, I. Ikeda, and T. Miyakawa. 2011. Long-term intake of fish oil increases oxidative stress and decreases lifespan in senescence-accelerated mice. *Nutrition* 27: 334–337.

Vanschoonbeek, K., M.A.H. Feijge, W.H.M. Saris, M.P.M. de Maat, and J.W.M. Heemskerk. 2007. Plasma triacyl-glycerol and coagulation factor concentrations predict the anticoagulant effect of dietary fish oil in overweight subjects. *The Journal of Nutrition* 137: 7–13.

Yang, B., T.G.P. Saldeen, W.W. Nichols, and J. Mehta. 1993. Dietary fish oil supplementation attenuates myocardial dysfunction and injury caused by global ischemia and reperfusion in isolated rat hearts. *The Journal of Nutrition* 123: 2067–2074.

Miscellaneous Foods and Food Components

9.1 Cereal Grains

These are excellent sources of digestible carbohydrates, dietary fiber, and proteins in addition to providing vitamins (B group and vitamin E) and minerals (zinc, phosphorus, selenium, and iron). Whole-grain cereals also contain high levels of bioactive phytochemicals such as phenolics, phytoestrogens, and antioxidants that could provide protective effects against human chronic diseases such as cancer, diabetes, kidney disease, and hypertension. In contrast to whole grains, refined grain products do not have the bran and germ parts of the seed, which causes reduced nutrient content or potential bioactive properties. One of the key earlier reported benefits of whole-grain consumption is the reduction in the risk of developing cardiovascular diseases, especially coronary heart disease (CHD) and stroke. The mechanism of action has been attributed to improved blood glucose levels, reduced body mass index (BMI), and increased insulin sensitivity. In fact, a 28% reduction in the risk of developing coronary heart disease (CHD) was found in people with highest consumption of whole grains when compared to those with lowest consumption. The fiber contents of whole grains contribute to delayed digestion and feeling of fullness, which may enhance glucose tolerance and removal of dietary lipids (especially cholesterol) in the feces. In the colon, the indigestible carbohydrates are fermented by microorganisms to form short-chain fatty acids (SCFAs) that inhibit carcinogenesis, increase acidity to destroy pathogens, and enhance removal of toxins, especially ammonia. Phenolics and phytoestrogens in whole grains can act as antioxidants that contribute to scavenging of toxic free radicals and reduce the potential for oxidant-induced DNA damage. In contrast, refined grains have high levels of readily digestible carbohydrates and low levels of fiber plus phenolics, which can increase blood lipid and glucose levels, cause insulin resistance, and lead to various cardiovascular disorders. Therefore, as part of a healthy diet, increase consumption of whole-grain food products is highly recommended.

9.1.1 Amaranth

On a dry weight basis, amaranth seeds contain approx. 14% total dietary fiber (6% soluble and 8% insoluble types), and 148 µg/g of polyphenols (mostly anthocyanins and flavonoids). In amaranth extracts, antioxidant activity is directly proportional to the level of polyphenols. Due to its contents of fiber and polyphenolics, amaranth could be used as a functional food that benefits the cardiovascular system. In rat experiments, addition of amaranth to a diet that contained excess cholesterol prevented the development of hypercholesterolemia accompanied by reductions

R.E. Aluko, *Functional Foods and Nutraceuticals*, Food Science Text Series,
DOI 10.1007/978-1-4614-3480-1_9, © Springer Science+Business Media, LLC 2012

in plasma LDL cholesterol and triglyceride levels. Therefore, amaranth may be used to formulate foods for hypercholesterolemic patients that are allergic to cereals.

9.1.2 Barley

Barley grains are rich sources of dietary fibers such as β-glucan, arabinoxylans, and cellulose. Consumption of diets rich in barley or β-glucan extracts is known to have beneficial physiological effects because of the soluble state and high molecular weight of this polysaccharide. β-glucan has been shown to increase daily fecal bile acids output, which leads to lower blood cholesterol and lipoprotein concentrations in human subjects. This is because in the gastrointestinal tract, β-glucan forms a viscous hydrated mass that traps bile acids to reduce the amount of cholesterol and other neutral sterols that are absorbed into the body. Dietary β-glucan also lowers postprandial blood glucose and insulin responses in humans. Ability to reduce blood glucose after a meal is because the viscous mass formed by β-glucan traps digestive enzymes, the substrates, and products, which leads to decreases in the rate of digestion of starch into glucose and the rate of absorption of glucose from the gastrointestinal tract into the bloodstream. Therefore, barley products in general and β-glucan in particular may be good for diabetics, obese, and hypercholesterolemic patients as a means of controlling level of blood sugar and cholesterol, respectively. Colonic fermentation of β-glucan and other barley dietary fiber components leads to the formation of SCFAs, which reduces intestinal pH and favors decreased formation of secondary metabolites from bile acids. This is because the drop in pH leads to an inhibition of the activity of 7α-dehydroxylase, a bacterial enzyme that converts bile acids into secondary metabolites. Secondary bile acids such deoxycholic and lithocholic acids are known promoters of colon cancer formation and pathogenesis. By suppressing formation of secondary bile acids and by decreasing transit time in the gut, high levels of dietary barley may also have a preventive role in colon cancer pathogenesis.

9.1.3 Wheat and Triticale

Wheat and triticale seeds are rich sources of dietary fibers in particular the arabinoxylans and hemicelluloses. Consumption of whole wheat and triticale flours leads to enhanced ceca (increased fecal bulk) and decreased bulk pH, which enhances the elimination of cholesterol and potential toxins or carcinogens. In rat experiments, addition of wheat or triticale flours to the diets led to significant increases in fecal total steroids (bile acids + sterols), which was accompanied by reductions in plasma triglyceride and cholesterol levels, but HDL cholesterol was not affected. This effect was more pronounced in the viscous wheat flour when compared to the less viscous wheat variety. There is also a marked increase in the SCFA pool size, mostly propionic and butyric acids especially when triticale is in the diet.

9.2 Flaxseed

Whole-grain flaxseed contains bioactive fatty acids and lignans among other nutrients. The fatty acids have been previously discussed in Chap. 2; therefore, the focus here will be on the major lignan called secoisolariciresinol diglucoside (SDG). It should be noted that SDG is also present in wheat and other grains, but flaxseed contains the highest level. The structure of SDG shown in Fig. 9.1 and various reports have indicated potential ability of this compound to inhibit tumor development in animal disease models. For example, SDG reduced

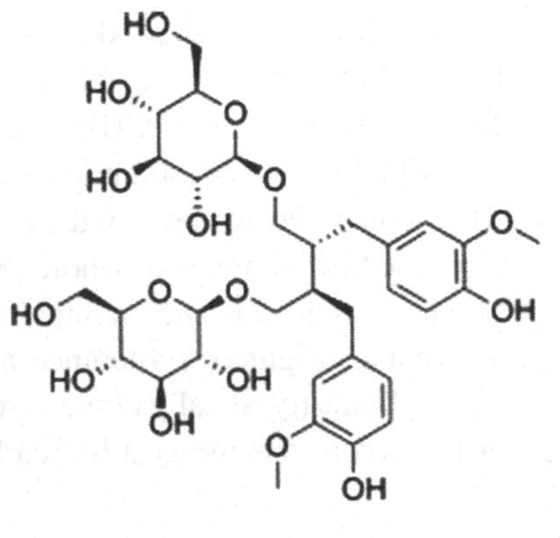

Fig. 9.1 Chemical structure of secoisolariciresinol diglucoside (SDG)

azoxymethane-induced formation of aberrant crypt foci in rats, while metastasis of melanoma cells was also reduced in mice. Bioavailability of SDG is very poor with reported ranges of a few nanomoles to a few micromoles per liter of human plasma or urine. Thus, the anticancer effect of SDG is believed to be mediated through its two main metabolites, enterolactone and enterodiol, that are formed via anaerobic fermentation of SDG by colon bacteria. Apart from anticancer effects of the metabolites, SDG in its native form can inhibit colon cancer cell growth through induction of S-phase cell cycle arrest, an effect that is similar to those of the metabolites. When tested in a cell culture system, SDG was more stable than its metabolites, which suggests that SDG may be more effective in providing longer-lasting effects.

9.3 Buckwheat

Not to be confused with wheat, buckwheats are more of a fruit or nut rather than being grasses or cereals. There are two major cultivated buckwheats: the common buckwheat, which is called *Fagopyrum esculentum*, and the bitter buckwheat, also known as tartary buckwheat (*F. tataricum* Gaertn). Both types are used as crops though cultivation of the common variety is greater than that of the bitter variety. Buckwheat seeds contain high levels of proteins and flavonoids that have been demonstrated to have bioactive properties in animal models of chronic diseases. Another compound called 2″-hydroxynicotianamine (HNA) has been shown to be present in the inner and middle layers of buckwheat seeds. HNA has strong in vitro inhibition of angiotensin-converting enzyme with IC_{50} (inhibitory concentration that reduced enzyme activity by 50%) value of 0.08 μM. However, the blood pressure lowering effect of HNA is yet to be demonstrated.

9.3.1 Cholesterol-Lowering Effect of Buckwheat Proteins

Common buckwheat protein (CBP) or tartary buckwheat protein (TBP) product can be produced by extracting the respective flours with an aqueous solution that has been adjusted to pH 8.0. After mixing thoroughly and centrifuged, a supernatant is obtained that is then adjusted to pH 4.5 using dilute acid to precipitate the major proteins present in the buckwheat flour. This process is referred to as isoelectric protein precipitation because majority of the buckwheat proteins have isoelectric point close to pH 4.5, where they are totally insoluble in aqueous solutions. The precipitated protein is then dispersed in water, neutralized to pH 7.0 using dilute alkaline solution, and then freeze-dried or spray-dried as the CBP or TBP. Rats fed cholesterol-rich diet that had been supplemented with CBP or TBP as the sole protein source had 25–32% reductions in serum cholesterol when compared to similar rats that had casein as the sole protein source. The degree of digestibility in the digestive tract is the main basis for the observed hypocholesterolemic effects of buckwheat proteins. In comparison to caseins, buckwheat proteins are less digestible when subjected to pepsin and pancreatin hydrolysis. Indigestible buckwheat proteins move into the colon and exhibit dietary fiber property by binding cholesterol, which prevent reabsorption of bile acids. Proof is in the fact that consumption of in vitro predigested buckwheat proteins was less effective in reducing serum cholesterol when compared to consumption of the intact proteins. Analogous to "resistant starch," the cholesterol-lowering ability of buckwheat proteins may be attributed to the fraction consisting of "resistant proteins." There is also a direct correlation between fecal nitrogen and neutral sterols, which further confirms that protein indigestibility contributes to expulsion of cholesterol from the digestive tract. Consumption of CBP and TBP was also associated with significant reductions in the *lithogenic index* though CBP was more effective than TBP. The lithogenic index is a measure of the amount of insoluble cholesterol, which could encourage formation of calculi (such as gallstones) in the body. The mechanism involved in the anti-lithogenic effects of CBP and TBP is believed to be through increased excretion of fecal bile neutral sterols, which prevents the bile from becoming excessively saturated with solids. The probability for gallstone or other forms of calculi formation reduces as the concentration of bile acids decreases. Therefore, CBP and TBP may be used as suitable ingredients to formulate functional foods or nutraceuticals with the potential to increase excretion of cholesterol and other neutral sterols. However, unlike dietary fibers, buckwheat proteins do not have the following effects:

1. Expulsion of acidic sterols
2. Increased weight of stool
3. Decreased colon pH

9.3.2 Common Buckwheat Proteins and Muscle Hypertrophy

In experiments conducted with rats, a common buckwheat protein extract (BWPE, 65.8% protein content) was found to increase muscle mass when compared with casein in the diet. At 10%, 20%, and 30% levels of addition of the BWPE, mass of the gastrocnemius muscle (g/kg body weight) increased by 10%, 15%, and 13%, respectively. For the plantaris muscle, up to 17% increase was obtained at 30% dietary BWPE while up to 14% increase in soleus muscle was obtained at the 20% BWPE level in comparison to similar levels of casein. Concomitant to muscle hypertrophy was a noticeable decrease in carcass fat when compared to casein-fed rats, which was attributed to decreased activity of lipogenesis enzymes in the BWPE-fed rats. It should be noted that the BWPE contained almost 3× the arginine content of casein. Arginine has been implicated in muscle protein synthesis, which could provide partial explanation for the observed muscle hypertrophy in BWPE-fed rats. Carcass protein and water were higher in the BWPE-fed rats, which taken together with the lower content of fat will suggest the formation of a leaner muscle mass than in the casein-fed rats. The contents of branched-chain amino acids (leucine, isoleucine, valine), which are associated with increased muscle protein synthesis, were similar in casein and BWPE; therefore, it is reasonable to suggest that the differences in muscle quality are due mostly to the higher arginine content of BWPE. It should be noted that apparent digestibility and hence in vivo amino acid availability of BWPE were

lower than that of casein; thus, the observed muscle hypertrophy could not have been due to better utilization of amino acids in BWPE. BWPE contains about 20% less content of protein, and the higher levels of nonprotein substances such as lipids and carbohydrate could have contributed to the observed muscle hypertrophy. Additional experiments are needed to define the roles of the nonprotein components in the hypertrophic effect of BWPE. Overall, BWPE may serve as a functional food that can induce increase muscle mass for increased strength, physical performance, and as an antiaging agent that can suppress the normal muscle loss that is associated with aging.

9.3.3 Effects of Common Buckwheat Solvent Extracts on Diabetes

An extract of common buckwheat seeds can be made by extraction of the ground bran and short fractions with aqueous ethanol solution followed by concentration by vacuum evaporation. Treatment of type 1 diabetic rat with this buckwheat extract produced up to 19% reduction in blood glucose level at 120 min after treatment. Treatment was effective only in the fed state and not in fasted diabetic rats. The buckwheat extract contained 0.2% level of D-*chiro*-inositol (D-CI), a compound that is believed to be the active component responsible for the blood glucose reduction effects. The mechanism for the antihyperglycemic effect of D-CI may involve inhibition of hepatic glucose output or increased glucose transport, in addition to increases in glucose utilization, disposal, and glycogen synthesis. However, though the exact mechanism has not been determined, buckwheat extract may serve as a therapeutic product in the management of type 1 diabetes mellitus. In addition to the use of extracts from regular seeds, sprouting has been used to improve the contents of flavonoid compounds in the seed. Following seed germination, buckwheat sprouts (BS) are produced by allowing the seeds to grow in the dark for 2 days and another 2 days under light. The seed sprouts are then removed, dried, and grinded to give the BS powder, which contains about 5% (w/w) of flavonoids (mainly rutin, orientin, isoorientin, vitexin, and isovitexin) in addition to about 0.08% (w/w) of anthocyanin (cyanidin 3-rutinoside). Oral administration of BS flavonoid compounds to mice led to reduced oxidative stress, which was demonstrated as improved lipid metabolism and inhibition of lipid peroxidation (measured as decreased plasma level of thiobarbituric acid reactive substances). Incorporation of BS extract into the diet of type 2 diabetic mice produced significant reductions in the plasma levels of glycated hemoglobin, plasma glucose, total cholesterol, HDL cholesterol, and arteriosclerotic index. These reductions in blood components were accompanied by significant reductions in liver and final body weights. The mechanism of action of the BS has been attributed to its ability to increase liver bile acid synthesis that enhances cholesterol removal through the feces; amount of fecal bile acids was higher in BS-fed mice. Moreover, the improved lipid metabolism may also be due to the fact that BS-fed mice had lower gene expression levels of fatty acid synthesis enzymes.

9.4   Tea

Most common teas are made from the leaves of the plant *Camellia sinensis* through drying and fermentation. Dry and unfermented leaves of *C. sinensis* are heated or steamed (to prevent fermentation), rolled and dried again to produce "green tea." Partially dried *C. sinensis* leaves are crushed and allowed to turn black through the process of fermentation and then dried to produce "black tea." Tea consists of mostly the polyphenols called *flavonoids* that can be as high as 150 mg in a 150-mL cup of brewed black tea. The most abundant tea polyphenols are the catechins, which are monomeric flavonoids. Epigallocatechin (EGC) and epigallocatechin gallate (EGCG) are the major catechins in tea, though differences abound in the levels found in green compared to black tea. EGCG accounts for 50–75% of the catechins in green tea, and therefore, most of the health benefits of this beverage are associated with the compound. The lower levels of catechins

in black tea are due to the fact that the leaves have been fermented during which there is increased polymerization of the monomeric catechins. In contrast, green tea is not fermented, and thus, most of the monomeric flavonoids are preserved. Strong antioxidative effects have been associated with tea flavonoids, though no substantial difference exists between green and black tea. Several potential health benefits have been associated with increased tea consumption such as the antimicrobial effects that can lead to improved gut and oral health in addition to resistance to infections. The prebiotic effects of tea flavonoids have also been associated with enhanced gut health through the elimination of pathogenic microorganisms in the feces. Tea flavonoids also have anti-inflammatory properties, which protect against arthritis and asthma. Bioavailability of tea catechins is low because of their short in vivo half-life (max. of about 5 h); therefore, repeated dosage throughout the day is an excellent way of maintaining high plasma levels. For example, oral administration of a cup of green tea can lead to an EGCG concentration of 1 μmol/l in the blood, while maximum human plasma level of 7.6 μmol/l can be achieved at a high oral dose of up to 23 mg/kg body weight. However, the 23 mg/kg body weight dose is eightfold greater than the daily intake from green tea and can only be achieved through ingestion of nutraceutical capsules that contain green tea extracts. Apart from flavonoids, tea also contains an amino acid called *theanine* that has been shown to enhance human innate immune functions, i.e., the compound improves natural immunity of the consumer so as to prevent entry of pathogenic agents or eliminate them if they enter the body. Theanine has been shown to be able to cross the blood–brain barrier, which may ameliorate cerebral endothelial damage and reduce the risk of stroke. The polyphenolic compounds in tea leaves have been associated with antioxidant protection in human subjects, while in tissue culture experiments, there was increased apoptosis of human cancer cells. Some of the specific health benefits of tea consumption are detailed below. Since some tea consumers use milk during the beverage preparation, it should be noted that milk proteins have been found to reduce increase energy expenditure (diet-induced thermogenesis, DIT) that is associated with tea polyphenols. The reduction in DIT could be attributed to protein-polyphenol interactions or coagulation of proteins in the acidic environment of the stomach. Both events could slow down emptying of the stomach and reduce the rate or extent of tea polyphenol metabolism. The negative effects of milk proteins are important because DIT has been proposed as part of the mechanism by which tea polyphenols enhance metabolic rate and attenuate weight gain, especially in obese conditions. Therefore, consumption of black tea or green tea without addition of milk may provide greater health benefits than tea that contains milk.

9.4.1 Tea Polyphenols and Alzheimer's Disease

EGCG has also been suggested as a potential agent that could reduce the development and severity of Alzheimer's disease (AD) because of its ability to inhibit aggregation of amyloid-β (Aβ), which is a major pathogenic factor in the disease. The neurotoxicity of Aβ was also reduced by EGCG treatment of neuronal cells, which may be related to direct disruption of Aβ fibril formation and breakup of already formed fibrils. Mechanism of fibril disruption is thought to involve prevention of fibrillogenesis as a result of hydrogen bonding of EGCG with the Aβ structure or covalent bonding with amyloid-like proteins. EGCG is known to bind and stabilize amyloidogenic proteins, which prevents formation of amorphous aggregates associated with fibril formation. EGCG is an activator of α-secretase, an enzyme that helps to reduce the rate at which Aβ peptide is generated in brain cells. EGCG may also act against AD because of its ability to scavenge reactive oxygen species and inhibit monoamine oxidase, which leads to reductions in peroxide levels. The anti-Aβ property has been demonstrated in vivo after oral feeding, which suggests bioavailability of EGCG.

9.4.2 Tea Polyphenols and Cardiovascular Diseases (CVD)

Compounds in green tea extract (GTE) may have substantial effects on the cardiovascular (CV) system through prevention of hypertension, cardiovascular damage, and endothelial dysfunction. In a rat model, it was demonstrated that GTE was able to prevent high dose angiotensin II-induced increases in blood pressure, left ventricular hypertrophy, and content of hydroperoxide radicals. Angiotensin II (Ang II) is a potent vasoconstrictor, and high levels normally lead to development of high blood pressure and associated hypertension symptoms. GTE was able to abolish the pressor effects of Ang II leading to lower blood pressure when compared to control where GTE was omitted. One of the mechanisms believed responsible for organ damage during high blood levels of Ang II is the high levels of superoxide dismutase (SOD-1), heme oxygenase 1 (HO-1), and NADPH oxidase endothelial p22^{phox} subunit, all leading to high levels of oxidative stress. The CV protective effects of GTE are due to its ability to scavenge superoxide free radicals as well as modulation of cellular functions that leads to reductions in the mRNA levels of HO-1, SOD-1, and p22^{phox} enzymes. The animal study was supported by an epidemiological finding in Japan which showed an inverse relationship between green tea consumption and cardiovascular mortality. People that drink more than five cups (approx. 500 ml) of green tea on a daily basis could have up to 33% reduction in the risk of CV mortality when compared to people that drink less than one cup daily, especially in patients with a history of arterial hypertension. A similar study in Norway showed that people that consume five cups or more of tea daily had lower systolic blood pressure and serum cholesterol (6–9 mg/dl reduction). In an Israeli study, a greater reduction of serum cholesterol (18–29 mg/dl) was observed for regular tea drinkers. Consumption of tea has been associated with decreased blood pressure, possibly mediated by ability of (−)-epicatechin to promote vasodilation through increased plasma levels of nitric oxide

(NO). The tea flavonoids have no measurable effect on the enzyme that catalyzes NO generation, endothelial nitric oxide synthase. Therefore, it is believed that the NO promoting effect of tea polyphenols is due to "NO-sparing" effect, i.e., preventing scavenging of NO by inhibiting NADPH oxidase and also through scavenging of toxic radicals that could inactivate NO.

Another plausible mechanism for the cardioprotective role of tea polyphenols is the ability to act as antioxidants that protect molecular integrity of LDL. Oxidized LDL will express chemotactic and adhesion molecules on the surface of endothelial cells. Additionally, the oxidized LDL molecules are scavenged by macrophages, which lead to the formation of lipid-laden foam cells that mark the onset of atherosclerotic lesions. By preventing LDL oxidation, it is possible to prevent or reduce the pathological intensity of atherosclerosis. When natural antioxidants that are present in the blood are overwhelmed, then the potential for LDL oxidation increases; therefore, supplementation of plasma antioxidants with similar compounds in the diet could provide therapeutic benefits. In human volunteers, intake of tea increased antioxidant capacity (AOC) of plasma in the hour following ingestion as well as increases in basal AOC after regular intake for up to 4 weeks. Increases in plasma AOC reflects higher plasma concentration of polyphenols from dietary sources and is an indication of the increased ability of the plasma to scavenge free radicals. It is possible that the higher AOC of plasma could have a sparing effect on LDL, especially copper-promoted oxidation. Daily consumption of green tea has been found to render LDL molecules resistant to in vivo oxidation either directly or through metabolites of the tea compounds. Green tea polyphenols also produce LDL sparing effect by lowering the rate of depletion of antioxidative vitamins. Daily consumption of tea provided substantial improvement in antioxidant status, which attenuated benzene-induced toxicity in humans exposed to benzene. Other human studies have also showed that regular tea consumption led to increased plasma glutathione, decreased plasma peroxides, and less DNA damage. However, a few human experiments have failed to show a convincing pattern of delayed LDL oxidation following ingestion of teas, though variations in testing methodology may be responsible for the conflicting results.

Tea polyphenols have also been associated with reduced risk of stroke as demonstrated through epidemiological and animal experiments. In various epidemiological reports, it has been shown that regular intake of various tea flavonoids reduced the risk of coronary heart disease mortality, ischemic stroke incidence cerebral infarction, intracerebral hemorrhage, stroke incidence, and stroke mortality in men and women. A 21% reduction in stroke risk has been shown to be associated with the consumption of each additional three cups of tea. In animal experiments, oral administration of EGCG, theanine, and theaflavin led to reductions in infarct volume, a measure of extent of blockage of the blood vessel. Tea flavonoids may also act as protective agents against neuron damage. In an animal model of stroke, intravenous administration of theaflavins (black flavonoids) resulted in protection of neurons against ischemia-reperfusion injury. In theaflavin-infused rats, there was attenuation of leukocyte infiltration as well as activities of inflammatory-related prooxidative enzymes. And green tea polyphenols can reduce the ischemia-reperfusion (IR)-induced injury to the intestinal mucosa. The injury occurs because IR enhances excess production of ROS (especially superoxide and H_2O_2) and inflammatory cytokines that overwhelm the natural antioxidant defense mechanisms (catalase, superoxide dismutase, glutathione peroxidase, etc.) in the mucosa. The IR injury can cause increased mucosal and vascular permeabilities to bacterial translocation, which can lead to systemic inflammation, respiratory failure, and multiple-organ failure. Because tea polyphenols can perform free radical scavenging functions, prior administration of green tea reduced level of inflammation and associated damage to the intestinal mucosa cells during IR.

9.4.3 Tea Polyphenols and Metabolic Syndrome

Obesity continues to increase at an alarming rate all over the world but more predominantly in the affluent Western countries. Associated with obesity are a group of risk factors that increase the chance for developing heart disease, diabetes, and stroke. The presence of these risk factors, which include large waist line, high blood pressure, high blood triglycerides, low HDL cholesterol, and high fasting blood sugar, constitutes the metabolic syndrome. Therefore, compounds that reduce these risk factors may be important in the management and prevention of diseases associated with metabolic syndrome. A green tea extract has been shown to improve glucose tolerance and insulin sensitivity in experimental mice. The beneficial effect on glucose metabolism is believed to be due to increased glucose transportation and lipid metabolism, as well as reduced expressions of gluconeogenic enzymes, such as phosphoenolpyruvate carboxykinase and glucose-6-phosphatase. Green tea catechins, especially EGCG, when incorporated into the diet of mice led to reductions in weight gain when compared to animals on control diet. EGCG is also an effective dietary agent for increasing fat oxidation and reducing brown adipose tissue as well as plasma insulin and glucose levels. Blood lipids, especially cholesterol, were reduced by EGCG, and this was reflected in decreased accumulation of lipids in the liver; intensity of fatty liver condition was decreased. The mechanism for reduced blood and liver lipids as well as loss of brown adipose tissue mass could be traced to the increased content of lipids in the feces, which suggests that EGCG acts through reduction in lipid digestion and/or absorption. Tea catechins inhibit pancreatic lipolytic enzymes such as lipases and phospholipase A2, which limit digestion of dietary fats. To prevent lipid absorption, tea catechins bind to dietary lipids, especially cholesterol, to alter entry into the lymphatic system. High oxidant or inflammatory status is also associated with metabolic syndrome, and EGCG treatment lowered the level of proinflammatory cytokines in a mice model of

the disease. Under obese conditions, there is upregulation of proinflammatory cytokines such as interleukin-6 (IL-6) and monocyte chemoattractant protein 1 (MCP-1), especially in the brown adipose tissue. These cytokines contribute to development of obesity-induced insulin resistance by reducing production of insulin-stimulated Akt/PKB phosphorylation. However, addition of EGCG to a high-fat diet led to reductions in the levels of MCP-1 and IL-6, which may have contributed to the observed decrease in insulin resistance. C-reactive protein (CRP) is an inflammatory risk factor that has been implicated in long-term cardiovascular morbidity, and obesity has been associated with high CRP levels during metabolic syndrome. Level of CRP was decreased in mice fed a diet that contained EGCG, suggesting potential protective effect of tea catechins against cardiovascular diseases. Tea catechins also have antiobesity effects through mechanisms that include direct inhibition of fat from the gut as indicated above, suppression of adipocyte differentiation and proliferation, and inhibition of the activity of catechol-o-methyl transferase (COMT). COMT is an enzyme present in brown adipose tissues where it suppresses fatty acid oxidation; therefore, inhibition of COMT by catechins promotes loss (oxidation) of fatty acids and reduced adipose tissue weight. EGCG has a negative effect on adipocytes by decreasing the levels of phosphorylated cellular proteins such as the ERK1/2, cdk2, and cyclin D1 that are required for cell division; this leads to arrest of cell growth. EGCG can also reduce adipose tissue mass by inducing apoptosis in mature adipocytes. Additional mechanism of the antiobesity activity of EGCG includes ability to increase expression and phosphorylation of adenosine monophosphate-activated protein kinase (AMPK) as well as acetyl-CoA carboxylase (ACC). AMPK is an important enzyme involved in the suppression of anabolic pathways while enhancing catabolic pathways. For example, AMPK phosphorylates ACC, which leads to reduction in the rate of fatty acid esterification to glycerides and in turn enhances oxidation of fatty acids. AMPK also directly inhibits hepatic HMG-CoA reductase, which leads to reduced cholesterol synthesis. Tea polyphenols have been shown to reduce the symptoms of type 2 diabetes through increasing glucose uptake by the skeletal muscles. The effect of tea polyphenols on skeletal muscle is mediated through upregulation of glucose transporter 4 (Glut 4), decreased translocation of Glut 4, and decreased levels of adipose tissue insulin. In human subcutaneous (but not visceral) preadipocytes, tea polyphenols reduced triglyceride incorporation into adipocytes during adipogenesis through suppression of adipocyte determination and differentiation factor 1 (ADD1), as well as other factors involved in adipocyte differentiation. Therefore, the antiobesity effects of tea polyphenols is also as a result of decreased adipogenesis coupled with increased lipolysis.

9.4.4 Tea Polyphenols and Cancer

In several animal experiments, oral administrations of black or green tea extracts were found to have protective effects against tumor initiation, promotion, and progression, though green tea seems to be more active. Epigallocatechin gallate (EGCG), which is the main polyphenolic compound in green tea and is also present at lower concentrations in black tea, has been identified as the main compound that is responsible for the anticancer properties of teas. For example, there was an inverse correlation between the intake levels of EGCG and number of lung tumors in experimental rats. However, in liver tumors where protective effect of tea was observed, there was no correlation between EGCG intake and number of tumors, which suggests that other polyphenols and nonpolyphenols in tea may also have chemoprotective properties. Caffeine may also play a significant role in the anticancer effects of teas. In experimental lung cancer, a caffeine solution was found to be as active as black tea that contained similar caffeine level. In UVB-induced carcinogenesis, decaffeinated green and black teas were found to be less protective than regular teas; activity was restored by addition of caffeine to the decaffeinated teas. Caffeine is not the only protective agent in teas since decaffeinated green and black teas

still exhibited significant anticarcinogenic and antimutagenic properties in several experiments. Apart from EGCG and caffeine, theaflavins (black tea polyphenol polymers) were found to have antimutagenic properties, inhibited transformation of a mouse epidermal cell line, and proliferation of a human epidermoid carcinoma cell line. Black tea theaflavins were also found to promote apoptosis of human lymphoma and stomach cancer cell lines. Oral administration of the theaflavins to rats inhibited both the esophageal and lung tumor formation.

Though both the tea polyphenols and caffeine are active anticancer agents, the route of administration has an effect on potency. For example, inhibition of skin cancer is greater with oral caffeine than oral tea polyphenols, whereas both agents have similar anticancer activity when applied topically. Thus, orally administered caffeine is able to reach the skin as a result of increased absorption into the blood, whereas other tea polyphenols are poorly absorbed from the GI tract. It is important to note that the lack of activity of orally administered tea polyphenols against skin cancer is not because of lack of inherent activity but because of decreased bioavailability. In contrast to the effect seen with skin cancer, tea polyphenols are highly more active against intestinal cancer when compared to caffeine. Even though they may have poor bioavailability, direct contact with the oral cavity and intestinal tract is probably responsible for the efficacy of tea polyphenols against cancers of the GI tract.

Tea polyphenols also inhibit cancer formation in lungs, prostate, and mammary glands, though bioavailability in these tissues is unknown. It is possible that the anticancer effects observed in these internal organ sites are due to polyphenol metabolites that are released into the blood or modulation of various hormone metabolic pathways. The tea polyphenol metabolites could act as inhibitors of intrinsic enzymes associated with several metabolic pathways that are involved in cell proliferation. Such enzymes include matrix metalloproteases, DNA methyl transferase, mitogen-activated protein kinases, phosphatidylinositol-3 kinase, and dihydrofolate reductase.

Modulation of these metabolic pathways by tea polyphenols could lead to anticancer activities such as increased apoptosis, prevention of tumor cell growth and proliferation, or reduction in rate of tissue angiogenesis (formation of new blood vessels). For example, work with green tea suggests that catechins act as angiogenesis inhibitors by blocking cellular receptor binding of angiogenesis-promoting agents such as angiogenin. Reduction in the binding efficiency of angiogenin (a 14-kDa protein that is a potent stimulator of blood vessel formation) will lead to reduced vascularization, which is not favorable for growth of cancerous tissues. Apoptosis induction in cancer cells was shown to be due to the prooxidant effects of EGCG, which was abolished when other antioxidants (catalase, glutathione, and ascorbic acid) were included in the treatment. Thus, the presence of EGCG led to increased production of free radicals that caused apoptosis of the cancer cells. A proposed mechanism of the inhibition of breast cancer cell growth by EGCG involves inhibition of nitric oxide (NO) and endothelial nitric oxide synthase (eNOS), the enzyme that converts arginine to NO. NO is a cell signaling molecule, and high levels have been found in tumor cells while tumor cell migration is dependent on eNOS activation. EGCG also reduced proliferation of breast cancer cells by inhibiting activity of guanylate cyclase and reducing the level of cGMP, a compound that promotes cell proliferation. Therefore, downregulation of eNOS activity and NO production by EGCG leads to reduced tumor growth and cell proliferation. Increased efficacy of tea polyphenols such as EGCG has been achieved through acetylation to the peracetylated derivative, which leads to increased absorption into the blood and greater apoptosis in human breast cancer cells. The anticancer effects of tea polyphenols may also be due to their ability to chelate iron, which will reduce the occurrence of Fenton reaction that generates toxic peroxides. Tea polyphenols may also inhibit phase I metabolizing enzymes such as cytochrome P450 and reduce conversion of inactive procarcinogens to active carcinogens. However, a note of caution that increased bioavailability of tea polyphenols could lead to undesirable levels

in tissues because high doses of EGCG can cause toxicity in the liver, kidneys, and intestine.

9.4.5 Tea Polyphenols and Food Digestion

It is a well-known fact that tea polyphenols can form complexes with proteins, which leads to precipitation and inactivation of enzymes. The interactions between polyphenols and enzymes occur mostly through noncovalent interactions; the galloyl (phenolic) and hydroxyl groups form hydrogen bonds with polar groups (amino, carboxyl, amide, guanidine, etc.) of proteins. Similarly, hydrophobic amino acid residues in enzyme proteins (leucine, isoleucine, proline, phenylalanine, tyrosine, valine) can form strong hydrophobic groups with tea phenolics. Therefore, interactions of enzymes with tea polyphenolics occur through hydrogen bonds and hydrophobic interactions, which distort enzyme conformation and lead to reductions in catalytic efficiency. This phenomenon has been demonstrated in vitro during which tea polyphenols inhibited catalytic activities of key digestive enzymes such as lipase, pepsin, α-amylase, and trypsin. Based on the inhibition of digestive enzymes, moderate consumption of tea may be used to limit food digestion and reduce the caloric value of diets, which could be beneficial for weight reduction. Or in the case of diabetes, prior consumption of tea may help reduce rate of digestion of carbohydrates and assist in the management of blood glucose.

9.4.6 Tea Polyphenols and Bone Health

Bone demineralization gradually leads to reduced bone mass and deterioration of the microarchitecture of bone tissue, which characterizes a degenerative bone disease called osteoporosis. Osteoporosis is basically a disease condition that results from an imbalance between activities of the bone-forming cells (osteoblasts) and those of the bone-resorbing cells (osteoclasts). As the disease progresses, the bone becomes more fragile

and is accompanied by increased risk or susceptibility to fractures of the wrist, hip, and spine.
Postmenopausal women have a higher risk of
developing osteoporosis because of decreased
estrogen levels, which leads to increased activity
of osteoclasts and reduced activities of osteoblasts.
Also, increased levels of reactive oxygen species
or free radicals coupled with decreased antioxidant status can lead to oxidative stress, which is
one of the causative factors for increased loss
(through apoptosis) of bone-forming cells.
Therefore, ingestion of antioxidants such as the
polyphenols found in tea could reduce bone
demineralization and enhance bone strength.
Some of the reported benefits of tea polyphenols
in bone health have been demonstrated mostly
through animal studies and include the following:

1. Reduced intensity of rheumatoid arthritis
   (RA), which is characterized by local bone
   erosion. ECGC can decrease bone demineralization, and treatment with EGCG led to
   reduced pathological intensity of RA.
   Therefore, tea polyphenols can inhibit activity
   of osteoclasts, the bone-resorbing cells whose
   excessive activity can cause reductions in the
   degree of bone mineralization.
2. Increased protection of bone cells against
   destructive oxygen radicals and lipid hydroper-
   oxides; tea polyphenols can chelate iron
   and prevent iron-dependent metal lipid
   peroxidation.
3. Increased bone mass density through reduction
   in oxidative stress, which leads to sparing of
   bone cells from estrogen deficiency-induced
   loss of antioxidant defenses (e.g., glutathione
   and catalases). This is because estrogen
   decreases oxidative stress in bone cells, and
   reductions in the level of this hormone in
   women can lead to high oxidative stress, which
   enhances destruction of bone-forming cells,
   especially osteoblasts. EGCG and ECG have
   also been shown to have weak binding affinity
   for estrogen receptors, and high plasma levels
   of the tea polyphenols could lead to stimulation
   of these receptors to mimic estrogen-induced
   stimulation of bone cell antioxidant defense.
   Increased antioxidant defense enhances activity of osteoblasts, the bone-forming cells.

9.4.7 Tea Polyphenols and Hepatic Injury

During injury, the loss of blood remains a leading
cause of death after trauma, and therefore, blood
resuscitation remains an important part of the
treatment regimen. However, during blood resuscitation, there is development of the systemic
inflammatory response syndrome (SIRS) that
causes metabolic dysfunction and can lead to
multiple-organ failure. This is because resuscitation can compromise the integrity of several
organs at the microcirculation level and cause
excessive activation of neutrophils with subsequent release of inflammatory cytokines. The
neutrophils cause hepatic damage through
increased production of ROS and reactive nitrogen species, which enhances lipid peroxidation-
induced damage to cellular biopolymers (proteins,
DNA, and membrane lipids). In rats that have
undergone hemorrhage/resuscitation (H/R), addition of green tea extract (GTE) to the diet led to
significantly reduced hepatic injury as measured
by necrosis and apoptosis. Hepatic oxidative
stress as measured by lipid oxidation and nitrosative stress as determined by 3-nitrotyrosine staining were significantly reduced by GTE. Plasma
content of alanine amino transferase (ALT) and
lactate dehydrogenase (LDH), which serve as
markers for hepatic cell and general cell damage,
respectively was each reduced significantly by
incorporation of GTE into the rat diet. These
beneficial effects of GTE are due to the ability of
constituent polyphenolic compounds to scavenge
free radicals and reduce oxidative stress. The
mechanism of action of GTE also involved suppression of a mitogen-activated protein kinase
called c-Jun NH2-terminal kinase (JNK). This
is because JNK activation enhances expression
of several inflammatory genes, production of
proinflammatory molecules such as IL-6, and
apoptosis subsequent to H/R and hypoxia. JNK is
also involved in proapoptotic and pronecrotic
pathways in tissues with high levels of oxidative
and nitrosative stress. Therefore, attenuation by
GTE of phosphorylation-mediated JNK activation contributed to the observed decrease in
level of hepatocyte apoptosis and necrosis when

compared to control rats. GTE in the diet led to decreased expression levels of NF-κB, a factor that is responsible for inducing expression of proinflammatory mediators that are involved in the pathogenesis of liver damage. Rats that consumed the GTE diet had reduced hepatic gene expression of Bcl-2-associated X protein (Bax) and caspase 8, both proapoptotic factors. Subsequent to H/R, the GTE diet enhanced hepatic gene expression of Bcl-2, the antiapoptotic factor. Bax normally competes with Bcl-2 and therefore, increased expression of the latter will favor reduced rate of apoptotic cell death. Thus, the use of GTE may provide therapeutic relief against hepatic injury during blood resuscitation.

9.5   Coffee and Caffeine

Coffee consumption has increased tremendously worldwide and is the second most important beverage after water with an estimated annual intake of 500 billion cups. Coffee is highly enriched in various phytochemicals of which caffeine is the most important. Coffee is the richest source of caffeine (~100 mg in 240 ml) and consumed mainly for the caffeine-induced stimulatory effects on the brain. It is estimated that ~70% of global caffeine consumption is from coffee. In addition to caffeine, other important phytochemicals in coffee include caffeic acid, chlorogenic acid, and hydroxyhydroquinone (Fig. 9.2). These coffee phytochemicals are potent antioxidants that could protect consumers against chronic diseases through their activities as scavengers of free radicals and modulators of cellular signaling pathways.

The enhanced sense of sensation and improved mental attention that are associated with coffee intake are due mainly to the action of the liver where the caffeine is metabolized by cytochrome P450 1A2 (CYP1A2) enzymes into various active metabolites such as paraxanthine, theobromine, theophylline, and dimethylxanthines. Once decaffeinated, the potential health benefits of coffee such as protection against cardiovascular diseases

(CVD), diabetes, Parkinson's, and Alzheimer's are due mainly to other phytochemicals, especially chlorogenic acid. While the mental stimulatory and antioxidant roles of coffee phytochemicals have been well characterized, there are conflicting data on health effects of specific compounds. For example, increased caffeine consumption has been shown to be a risk factor for breast and prostate cancer, as well as development of rheumatoid arthritis and osteoporosis. Coffee phytochemicals such as cafestol and kahweol are linked to the incidence of hypercholesterolemia but may also have chemopreventive effects against certain diseases.

Melanoidins (high-molecular-weight nitrogen-containing polymers) are produced from Maillard reactions that occur during roasting of coffee beans. These melanoidins have been shown to have in vitro inhibitory effects against zinc-containing matrix metalloproteinases (MMPs), a group of enzymes that have shown to be involved in the pathogenesis of various inflammatory diseases and are essential promoters of cancer metastasis. The MMP inhibition by coffee melanoidins (>10 kDa) was shown to be independent of zinc concentration, which suggests that interactions with the enzyme protein are responsible for reduced enzyme activity. Since polymers from green (unroasted) beans had no MMP inhibition, it follows that roasting-induced melanoidins were responsible for the MMP inhibition. But the potential for use of melanoidin-enriched coffee to prevent cancer pathogenesis will need to be confirmed using in vivo methods.

9.5.1   Caffeine and Diabetes

Using a high-fat diet-induced mice model of diabetes, daily oral administration of caffeine resulted in improved glucose tolerance, insulin sensitivity, and hyperinsulinemia when compared to mice that were given pure water. By improving insulin resistance, caffeine was able to suppress the high-fat diet-induced impaired glucose tolerance. Food intake and body weight were not affected by caffeine treatment, which indicates

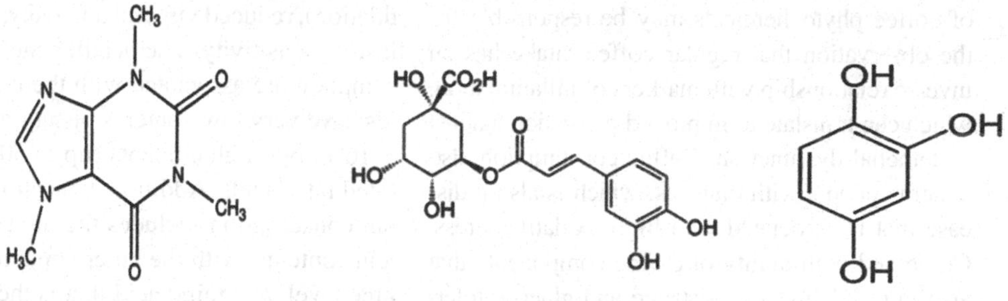

Fig. 9.2 Chemical structure of important coffee phytochemicals

that observed differences in measured health parameters were due to caffeine and not as a result of weight loss. The caffeine-treated mice also had reduced expression of adipose tissue mRNA for proinflammatory cytokines such as monocyte chemoattractant protein 1 (MCP-1) and interleukin-6 (IL-6). MCP-1 is responsible for enhancing insulin resistance in peripheral tissues through tumor necrosis factor alpha-induced increased phosphorylation (serine or threonine residues) of insulin receptor substrate-1 (IRS-1). Phosphorylated IRS-1 becomes insensitive to insulin-stimulated tyrosine phosphorylation, which leads to development of insulin resistance. IL-6 upregulates expression of the suppressor of cytokine signaling (SOCS) by activating adipose tissue and hepatic Janus kinase-signal transduction pathway. High SOCS levels lead to inhibition of insulin signaling and development of insulin resistance. Therefore, lower levels of MCP-1 and IL-6 in caffeine-fed mice suggest a potential role for caffeine in the management of type 2 diabetes. There was attenuated hepatic fatty acid accumulation as evident in the reduced expression of mRNA for genes related to fatty acid synthesis in mice that received caffeine. Specifically, the caffeine group had reduced levels of sterol regulatory element-binding protein-1 (SREBP-1), fatty acid synthase (FAS), and acetyl-CoA carboxylase (ACC). SREBP-1 is a known activator of FAS and ACC, which leads to increased fatty acid synthesis. High levels of fatty acids and lipid metabolites in the liver can interfere with insulin-stimulated tyrosine phosphorylation of IRS-1 and IRS-2, which leads to development of insulin resistance.

9.5.2 Coffee and Brain Disorders

Recent epidemiological study showed an inverse relationship between caffeine intake and the risk of Alzheimer's disease. The mechanism of action also involves caffeine, which is believed to induce reduced expressions of presenilin 1 β-secretase, enzymes that play important role in enhancing amyloid formation in the brain. Regeneration of brain neurons, especially dendrites and axons, has been shown to be promoted by trigonelline, an important alkaloid present in coffee. Since brain disorders are often associated with tissue damage or neurodegeneration, regular consumption of coffee may help repair these damages and limit the risk of brain disorders. Through their antioxidant properties, coffee constituents may promote increase scavenging of brain tissue-damaging free radicals and prevent tissue degeneration. The free radical and antioxidant effects of coffee phytochemicals have also been suggested as potential mechanism by which regular consumption of coffee limits development and pathological progression of Parkinson's disease.

9.5.3 Coffee and Cardiovascular Diseases

The role of coffee in preventing the incidence of CVD remains controversial, but there are epidemiological and basic research data that suggest potential benefits. For example, a cohort study showed an inverse relationship between regular coffee consumption and total mortality from coronary heart disease. It is believed that the antioxidant capacity of coffee phytochemicals may be responsible for the observation that regular coffee intake has an inverse relationship with markers of inflammation, which can translate to improved protection against endothelial dysfunction. Coffee consumption also benefits people with diabetes, which is also a disease that is associated with high oxidative stress. Combined with ability of coffee components that also improve insulin resistance and glucose tolerance, the free radical scavenging effects of coffee phytochemicals may help reduce CVD-related mortality in diabetic patients. Regular consumption of coffee has also been shown to reduce coronary calcification in women, which reduces the risk for atherosclerosis and myocardial infarction. This effect may be due to the fact that regular coffee intake reduced susceptibility of LDL to oxidative damage and production of lipid peroxides, which is attributable to the antioxidant properties of coffee phytochemicals. It has been shown that incorporation of coffee constituents such as p-coumaric, caffeic, and ferulic acids into LDL increases significantly following coffee consumption. The co-location of these coffee antioxidants with LDL will offer protection against oxidative stress-induced lipid peroxidation. In healthy men, long-term coffee consumption led to reduced platelet activation and plasma level of C-reactive protein; these effects are beneficial for improved cardiovascular health. However, it should be noted that excess daily (>8 cups/day) intake of coffee may have adverse effects in promoting CVD due to aggravation of cardiac arrhythmias and plasma homocysteine. Overall, excessive intake of coffee alone may not be a sufficient aggravating factor in promoting pathogenesis of CVD, except in association with other risk factors such as smoking, alcohol consumption, and plasma level of cholesterol.

9.6   Plant Nuts

Nuts have very high lipid contents and are energy dense, but regular consumption does not contribute to the incidence of chronic diseases such as cardiovascular, obesity, and diabetes. For example, frequency of nut consumption is associated with improved cardiovascular functions (flow-mediated dilation), reduced visceral adiposity, and improved insulin sensitivity. The health benefits of nut consumption are associated with the fact that nut lipids have very low contents of saturated fatty acids (<16%) but high contents (up to 50%) of unsaturated fatty acids. Additional nutritional benefits of nut consumption includes the high fiber and protein contents, with the latter containing high arginine level, an amino acid that is the precursor for synthesis of the vasodilatory compound, nitric oxide. Nuts are good sources of antioxidants (tocopherols and phytosterols), and micronutrients, especially folate that is important for detoxification of homocysteine, a pro-atherothrombotic amino acid. Phytosterols play a vital role in maintaining the structural integrity of phospholipid bilayers of nut cell membranes, and their molecular similarity to cholesterol makes nut consumption an important tool in maintaining cardiovascular health. By competing with and reducing cholesterol absorption, nut phytosterols can help to reduce the incidence of cholesterolemia and prevent negative cardiovascular events. Various epidemiological studies have shown that significant reductions in the risk of death from coronary heart disease are associated with regular consumption of nuts. The most reductions in risk were seen in people who consume nuts more than five times a week, but interestingly, even modest reductions were also associated with 1–4 portions per month. Similarly, the risk of diabetes was also inversely correlated to frequency of nut consumption with >7 intakes per week being the most effective, while modest risk reductions were seen with 1–6 intakes per week. Overall, each weekly serving of nuts was associated with an average of 8.3 reductions in the incident of coronary heart disease. The mechanism of action seems to be associated with the fact that frequency of nut consumption is inversely correlated to the levels of inflammatory markers such as IL-6, C-reactive protein (CRP), and fibrinogen. Plasma levels of adiponectin, an adipose tissue-secreted cytokine with anti-inflammatory and antiatherosclerotic properties, were also shown to be positively correlated with frequency of nut consumption. Levels of intercellular adhesion molecule-1 (ICAM-1) were inversely correlated with frequency of nut consumption. Frequency of nut consumption was negatively related to the incidence of gallstone disease with about 25% reduction in the risk for populations that have frequency of nut consumption >5/week.

Initial human intervention study was carried out at the Loma Linda University (reported in 1993), and it examined the effects of walnut consumption on blood lipid profile. Their results showed significant 12% and 18% reductions in plasma cholesterol and LDL cholesterol, respectively, of healthy subjects. Since the 1993 report, various studies have also shown that nut consumption produced dose-dependent reductions in plasma total and LDL cholesterol; these beneficial effects were independent of type of nut consumed. However, the negative effects of nut consumption on plasma total cholesterol and LDL cholesterol were greater for people with higher initial values and those with lower body mass index. But, it should be noted that the health benefits of nut consumption were not observed in people with metabolic syndrome, probably because of the altered cholesterol metabolism often seen in obese and insulin-resistant patients. In the insulin-resistant state, there is a high rate of hepatic cholesterol synthesis, which downregulates LDL receptors and makes them resistant to regulation by changes in dietary fatty acids. The low rate of intestinal cholesterol absorption produces a dampening effect on the cholesterol-raising response to dietary cholesterol as well as the cholesterol-lowering effects of plant sterols. Since one of the mechanisms by which nuts can reduce plasma cholesterol is through phytosterol-dependent lowering of cholesterol absorption, the dampening effect reduces the phytosterol flux through enterocytes, which in turn lowers the competitive level required to prevent cholesterol absorption. Even though plasma cholesterol levels are not affected by nut consumption in people with metabolic syndrome, there is evidence of improved insulin sensitivity and fasting glucose levels. Apart from reducing plasma cholesterol levels, nut consumption has beneficial effects in the prevention of lipid (LDL) oxidation. It is well known that oxidized LDL particles attract macrophages and cholesterol deposition

and are important disease enhancers in the pathogenesis of atherosclerosis. Consumption of nuts that are particularly rich in monounsaturated fatty acids (MUFA) has been shown to be beneficial in reducing the incidence of LDL oxidation. This is because MUFA is not a substrate for lipid oxidation, and their incorporation into LDL particles will decrease susceptibility to oxidation. But nuts also contain high contents of polyunsaturated fatty acids (PUFA), and therefore, there is potential for the occurrence of detrimental lipid oxidation. However, the presence of natural antioxidant compounds in these nuts could help suppress the rate of PUFA oxidation following nut consumption. This has been confirmed in various human intervention studies that showed absence of increased oxidative stress during nut consumption when compared to nut-free diets. In fact, few human studies have actually showed beneficial effects of nut consumption in reducing postprandial oxidative stress, while no study has shown negative effects. A major concern with nut consumption is allergic reaction to the major storage proteins, and therefore, avoidance must be used to prevent occurrence of adverse and potentially fatal allergic response. The bioactive roles of nuts can be summarized as follows.

Role of nuts in cardiovascular disease prevention:

- Nuts contain mainly unsaturated and polyunsaturated fatty acids.
- Walnuts, for example, are rich in ellagic acid, which has several cardioprotective effects.
- Good source of fiber.
- Recent survey of 10,000 men and 16,700 women showed that the frequency of nut consumption had an inverse relationship with risk of myocardial infarction or dying of coronary heart disease.

Possible mechanism of action:

- Improvements in serum lipid profiles, especially decrease in LDL and cholesterol levels.
- Nut proteins are high in arginine – precursor of nitric oxide (NO), a potent endogenous vasodilator.
- The high content of ellagic acid protects cardiac mitochondria against oxidative stress and reduces the potential for free radical-mediated cell death.

- Antiatherogenic effects:
  - NO can induce relaxation of vascular smooth muscle.
  - NO can inhibit platelet aggregation, monocyte adherence, and vascular smooth muscle cell proliferation.
  - The dietary fiber of nuts contains as much as 25% soluble fiber, which benefits lower total and LDL cholesterol levels.
  - Except vegetable oils, nuts have the highest levels of tocopherols (vitamin E), which protect cells against oxidative damage from LDL (unsaturated fatty acids).
  - Oxidized LDL has been shown to be cytotoxic and is also believed to induce cholesterol accumulation in blood vessels.
  - Folic acid in nuts may decrease blood homocysteine levels. Elevated homocysteine concentration is a risk factor for CHD.
  - Nuts contain approximately 8–20% of the daily recommended intake of magnesium, an element that improves heart function and necessary to maintain normal blood pressure.
  - Nuts contain 18% of the daily recommended intake of copper. Copper plays a key role in hematopoiesis (production of blood cells and platelets in the bone marrow).
  - Nuts contain many phytochemicals (flavonoids, phenolic compounds, isoflavones, sterols, etc.), which have been shown to be inversely related to CHD.

9.7   Mushrooms

Mushrooms play an important role as sources of nutrients in the diet of several people in the world, especially those living in Asian countries such as China and Japan. But mushrooms also have a history of use as medicinal resources for treatment of various illnesses. In general, a mushroom is the fruiting body of a macrofungus, which is manifested as an aerial umbrella-shaped fleshy growth that is large enough to be seen by the naked eye and picked by hand. Initial work in the late fifties showed that mushrooms belonging to the class of *Basidiomycetes* showed the presence of a substance that inhibited growth of sarcoma S180 tumor cells. Subsequent works showed that some types of polysaccharides in various mushrooms had antitumor properties when evaluated in rodent models of cancer, including adenocarcinoma and leukemia. The mushroom bioactive polysaccharides exist in different forms, from homopolymers (consisting of a single type of monosaccharide repeating unit) and heteropolymers (different types of monosaccharide repeating units) to polysaccharide-protein and polysaccharide-peptide conjugates. The mushroom polysaccharides could be in the form of linear or branched configuration with several types of glycosidic linkages: mostly α-1,4; β-1,3; β-1,6; and β-1,2.

9.8   Honey

After feeding on flower nectars, certain bees (genus *Apis*) produce honey (basically regurgitated nectar) that is stored as food for the insect colony. Honey is a sweet semisolid product with complex chemical composition. The sweetness of honey is due mainly to the high levels of fructose, which is the primary sugar. Honey also contains glucose and water as primary constituents in addition to other compounds such as nondigestible oligosaccharides, vitamins, minerals, and various antioxidants. One of the first reported effects of honey on human health is regulation of body weight. Animal experiments have shown that honey-based diets can reduce body weight in mature animals or body weight gain in growing animals when compared to sucrose-based diet. Moreover, the antioxidant properties of honey can reduce lipid peroxidation and contribute to increased cardiovascular health. For example, honey-based diets reduced hypertriglyceridemic effect that is normally associated with high-fructose diet. The high antioxidant property of honey reduced lipid peroxidation and associated build of vascular lipids (atherosclerosis). Antioxidant capacity of honey may be related to its source since darker honeys will have higher polyphenolic contents than lighter honeys. Reduced weight gain in growing animals was traced to reduction in food intake, which was attributed to altered production and increased sensitivity to leptin (appetite-suppressing hormone). Sucrose-fed rats have high levels of circulating leptin but develop leptin resistance, which promotes excessive food consumption that leads to increased/undesirable weight gain. Specifically, the reduced weight of honey-fed rats was correlated with reduced levels of plasma triglycerides and lower degree of adiposity, which is beneficial for people with diabetes. The lower plasma triglycerides associated with honey consumption are believed to be partly due to the presence of other sugars such as fructooligosaccharides (FOS) and isomaltulose, which are present as minor constituents. These minor oligosaccharides are nondigestible in the upper tract and become fermented into short-chain fatty acids (butyrate, isobutyrate, propionate) in the colon where they alter intestinal microflora towards beneficial lipid metabolism. The oligosaccharides are also known to inhibit lipogenic gene expression in experimental rats and reduce fatty acid synthesis in the liver, which would lead to reduced plasma triglyceride levels. However, it should be noted that most of the effects associated with FOS come from experiments that use feed levels that are relatively higher than FOS levels in honey. Therefore, whether the low levels of FOS in honey can have proposed health benefits remains to be confirmed.

9.9 Plant Protein Products

Protein-enriched products from plant seeds can be in the form concentrates (60–80% protein content) or isolates (>80%). These products represent cheap sources of bioactive compounds that can be used in the prevention and treatment of chronic diseases. This is due to the high protein and low fat contents of these products in addition to high fiber content of some of them. In rats that were fed sesame seed protein isolate, plasma lipid peroxidation was reduced up to 64% when compared to casein. There were simultaneous reductions in the plasma total cholesterol, LDL cholesterol, and triglycerides, whereas HDL

cholesterol was increased in the sesame protein-fed groups when compared to casein. The plasma cholesterol-lowering effects of plant proteins have been attributed to their low lysine/arginine ratio usually (<1.0) when compared to animal proteins like casein (usually >1.5). The proposed mechanism is that lysine is an inhibitor of arginase that breaks down arginine; therefore, high plasma levels of lysine enhance availability of arginine for incorporation into an atherogenic apoprotein. Low lysine/arginine ratios enhance arginine breakdown and reduced synthesis of arginine-rich atherogenic apoproteins. Sesame seed protein isolate has a very low (0.22) lysine/arginine ratio, which may have contributed to the observed decreases in plasma lipid indicators of atherogenesis. More importantly, rats on the sesame seed protein diet had reduced (up to 68%) levels of red blood cell membrane lipid peroxidation and up to 76% reduction in liver lipid peroxidation when compared to the casein diet. The mechanism behind the antioxidative effects of plant proteins is not completely understood, but data from the sesame seed protein isolate work suggest that high methionine levels could be beneficial for synthesis of antioxidant enzymes.

The hypolipidemic properties of pea (*Pisum sativum*) proteins have also been demonstrated through rat feeding experiments. In growing rats, incorporation of pea protein isolate into the diet led to significant reductions in plasma total cholesterol and triglyceride concentrations when compared with casein-based diet. The hypolipidemic properties of pea proteins were associated with increased LDL receptor mRNA expression, which indicates enhanced ability of the hepatic cells to remove LDL cholesterol from the blood circulatory system. Thus, increased hepatic catabolism of LDL particles could be responsible for the reduced plasma LDL cholesterol level associated with the pea protein diet. Pea protein-fed rats also had reduced expressions of mRNA for fatty acid synthase and stearoyl-CoA desaturase when compared to the casein-fed rats. Thus, reduced fatty acid synthesis contributed to lower plasma lipid levels in the pea protein-fed rats. The hypolipidemic effects of pea proteins may also be due to the relatively high arginine/lysine ratio (in comparison to the casein diet) as already noted above for sesame proteins.

9.10 Cocoa and Chocolate Products

Historically, cocoa (*Theobroma cacao*) was used as a medicinal product in the treatment of several diseases, but such uses have virtually disappeared with the advent of modern civilization. Cocoa beans and hence the chocolate products made from them have very high levels of flavonoids, specifically the subclass called flavanols. Recent evidence from scientific literature suggests that this group of polyphenols is responsible for the health-promoting effects such as increased vascular relaxation and reduced risk of cardiovascular mortality associated with consumption of cocoa products. It has been estimated that cocoa bean contains up to 6–8% (dry weight basis) of total polyphenols. It should be noted that the unprocessed cacao is different from the processed chocolate because the latter contains added ingredients, especially sugars in addition to the cocoa base. Moreover, the flavanol content of cocoa and chocolate products will depend on cultivar type, postharvest handling, and processing methods. Initial evidence supporting the health-promoting effects of cocoa product came from an epidemiological study of the native Kuna Indians that live in the San Blas Islands off the coast of Panama. The Kuna Indians are known to consume highly salty foods and drink several servings of unprocessed cocoa per day, but are known to be largely free of hypertension and age-related increases in blood pressure. The incidence of cardiovascular diseases, diabetes, and cancer-related mortality and morbidity is also very low in the Kuna Indian population. Further studies showed that European men that consume high levels of cocoa products had mean systolic (SBP) and diastolic (DBP) blood pressure that was 3.7 and 2.1 mmHg, respectively, less than men who consume lower amounts during a 15-year follow-up study. In a short-term study involving healthy and hypertensive human subjects, consumption of dark chocolate for 15 days led to significant reductions in SBP (−3.82 mmHg) and DBP (−3.92 mmHg) in addition to improved NO-dependent flow-mediated dilation. Such blood pressure-reducing effects were not observed when flavanol-free chocolate was used for the clinical trial. These reductions are significant because a 3-mmHg reduction in SBP can lead to an 8% reduction in the risk of stroke mortality, 5% reduction in coronary artery disease mortality, and 4% reduction of all-cause mortality. Overall, high level of chocolate consumption had an inverse relationship with cardiac mortality. The mechanism involved in the cardiovascular health benefits of cocoa products is believed to be associated with modulation of the renin-angiotensin system and NO production. Specifically, cocoa flavanols and procyanidins inhibited activity of angiotensin-converting enzyme (ACE), an enzyme that forms angiotensin II, a powerful vasopressor. Angiotensin II is also a prooxidant because it activates NAD(P)H oxidase, enzymes that produces reactive oxygen species. Therefore, cocoa polyphenols act to lower blood pressure by inhibiting ACE activity, which decreases level of angiotensin II and reduces NAD(P)H oxidase activation. The bioavailability of cocoa polyphenols has been confirmed from human trials which showed presence in the plasma following consumption of flavanol-rich cocoa drinks. In animal experiments, it has been demonstrated that ACE inhibition is associated with enhanced plasma levels of NO and reduced oxidative stress.

A cocoa fiber product derived from β-glucanase-catalyzed breakdown of cocoa husk was shown to have potential cardiovascular benefits because it reduced blood pressure in spontaneously hypertensive rats. The soluble cocoa fiber (SCF) product had ~41% soluble dietary fiber content and ~2.2% level of polyphenolic compounds. After 17 weeks of feeding, the SCF had no detrimental effect on body weight of the rats. SBP was significantly decreased by the SCF intake throughout the treatment period; however, DBP decreased only until week 14 and then was the same as the control for the remainder of the experiment. Treatment with the SCF was associated with slight decreases in plasma ACE activity and increased acetylcholine-induced vasorelaxation, both of which could have contributed to the observed decreases in blood pressure. Polyphenols have been shown to have ACE-inhibitory properties, which would explain the observed decreases in plasma ACE activity in rats that received SCF treatment. Since dietary fibers have been shown to attenuate severity of insulin resistance, which exists in SHR, it is possible that improved insulin management may have contributed additionally in reducing blood pressure. There was also significant decrease in lipid peroxidation as indicated by the reduced level of malondialdehyde (MDA) in the rats that ingested the SCF. The reduced MDA level could have been due to the antioxidant properties of polyphenols present in the SCF, which would have limited free radical-mediated lipid peroxidation. By reducing lipid peroxidation, the SCF would have contributed to reduced risk of atherosclerosis formation and hence the better vasorelaxative properties when compared to water. Thus, the cardiovascular benefits of SCF were due to the contents of soluble fiber as well as the presence of polyphenols, which enhanced vasorelaxation, lowered ACE activity and led to reductions in blood pressure.

Bibliography

Adam, A., M.-A. Levrat-Verny, H.W. Lopez, M. Leuillet, C. Demigne, and C. Remesy. 2001. Whole wheat and triticale flours with differing viscosities stimulate cecal fermentations and lower plasma and hepatic lipids in rats. *The Journal of Nutrition* 131: 1770–1776.

Antonello, M., D. Montemurro, M. Bolognesi, M. Di Pascoli, A. Piva, F. Grego, D. Sticchi, L. Giuliani, S. Garbisa, and G. Paolo Rossi. 2007. Prevention of hypertension, cardiovascular damage and endothelial dysfunction with green tea extracts. *American Journal of Hypertension* 20: 1321–1328.

Ayella, A., S. Lim, Y. Jiang, T. Iwamoto, D. Lin, J. Tomich, and W. Wang. 2010. Cytostatic inhibition of cancer cell growth by lignan secoisolariciresinol diglucoside. *Nutrition Research* 30: 762–769.

Biswas, A., P. Dhar, and S. Ghosh. 2010. Antihyperlipidemic effect of sesame (*Sesamum indicum* L.) protein isolate in rats fed a normal and high cholesterol diet. *Journal of Food Science* 75: H274–H279.

Butt, M.S., and M.T. Sultan. 2011. Coffee and its consumption: benefits and risks. *Critical Reviews in Food Science and Nutrition* 51: 363–373.

Chen, Y.-K., C. Cheung, K.R. Reuhl, A.B. Liu, M.-J. Lee, Y. P. Lu, and C.S. Yang. 2011. Effects of green tea polyphenol (−)-epigallocatechin-3-gallate on newly developed high-fat/western-style diet-induced obesity

and metabolic syndrome in mice. *Journal of Agricultural and Food Chemistry* 59: 11862–11871.

Czerwinski, J., E. Bartnikowska, H. Leontowicz, E. Lange, M. Leontowicz, S. Katrich, S. Trakhtenberg, and S. Gorinstein. 2004. Oat (*Avena sativa* L.) and amaranth (*Amaranthus hypochondriacus*) meals positively affect plasma lipid profile in rats fed cholesterol-containing diets. *The Journal of Nutritional Biochemistry* 15: 622–629.

De Marco, L.M., S. Fischer, and T. Henle. 2011. High molecular weight coffee melanoidins are inhibitors for matrix metalloproteinases. *Journal of Agricultural and Food Chemistry* 59: 11417–11423.

Dongowski, G., M. Huth, E. Gebhardt, and W. Flamme. 2002. Dietary fiber-rich barley products beneficially affect the intestinal tract of rats. *The Journal of Nutrition* 132: 3704–3714.

Flight, I., and P. Clifton. 2006. Cereal grains and legumes in the prevention of coronary heart disease and stroke: a review of the literature. *European Journal of Clinical Nutrition* 60: 1145–1159.

Grassi, D., G. Desideri, and C. Ferri. 2010. Blood pressure and cardiovascular risk: what about cocoa and chocolate? *Archives of Biochemistry and Biophysics* 501: 112–115.

Hamer, M. 2007. The beneficial effects of tea on immune function and inflammation: a review of evidence from in vitro, animal and human research. *Nutrition Research* 27: 373–379.

Harvey, B.S., I.F. Musgrave, K.S. Ohlsson, A. Fransson, and S.D. Smid. 2011. The green tea polyphenol (−)-epigallocatechin-3-gallate inhibits amyloid-β evoked fibril formation and neuronal cell death in vitro. *Food Chemistry* 129: 1729–1736.

He, Q., Y. Lv, and K. Yao. 2006. Effects of tea polyphenols on the activities of α-amylase, pepsin, trypsin and lipase. *Food Chemistry* 101: 1178–1182.

Kanti Maiti, T., J. Chatterjee, and S. Dasgupta. 2003. Effect of green tea polyphenols on angiogenesis induced by an angiogenin-like protein. *Biochemical and Biophysical Research Communications* 308: 64–67.

Kawa, J.M., C.G. Taylor, and R. Przybylski. 2003. Buckwheat concentrate reduces serum glucose in streptozotocin-diabetic rats. *Journal of Agricultural and Food Chemistry* 51: 7287–7291.

Kayashita, J., I. Shimaoka, M. Nakajoh, M. Yamazaki, and N. Kato. 1997. Consumption of buckwheat protein lowers plasma cholesterol and raises fecal neutral sterols in cholesterol-fed rats because of its low digestibility. *The Journal of Nutrition* 127: 1395–1400.

Kayashita, J., I. Shimaoka, M. Nakajoh, M. Kondoh, K. Hayashi, and N. Kato. 1999. Muscle hypertrophy in rats fed on a buckwheat protein extract. *Bioscience, Biotechnology, and Biochemistry* 63: 1242–1245.

Matsuda, Y., M. Kobayashi, R. Yamauchi, M. Ojika, M. Hiramitsu, T. Inoue, T. Katagiri, A. Murai, and F. Horio. 2011. Coffee and caffeine improve insulin sensitivity and glucose tolerance in C57BL/6J mice fed a high-fat diet. *Bioscience, Biotechnology, and Biochemistry* 75: 2309–23015.

Miura, Y., T. Chiba, S. Miura, I. Tomita, K. Umegaki, M. Ikeda, and T. Tomita. 2000. Green tea polyphenols (flavan 3-ols) prevent oxidative modification of low density lipoproteins: an ex vivo study in humans. *The Journal of Nutritional Biochemistry* 11: 216–222.

Nemoseck, T.M., E.G. Carmody, A. Furhner-Evanson, M. Gleason, A. Li, H. Potter, L.M. Rezende, K.L. Lane, and M. Kern. 2011. Honey promotes lower weight gain, adiposity, and triglycerides than sucrose in rats. *Nutrition Research* 31: 55–60.

Relja, B., E. Tottel, L. Breig, D. Henrich, H. Schneider, I. Marzi, and M. Lehnert. 2012. Plant polyphenols attenuate hepatic injury after hemorrhage/resuscitation by inhibition of apoptosis, oxidative stress, and inflammation via NF-kappaB in rats. *European Journal of Nutrition* 51: 311–321.

Rigamonti, E., C. Parolini, M. Marchesi, E. Diani, S. Brambilla, C.R. Sirtoli, and G. Chiesa. 2010. Hypolipidemic effect of dietary pea proteins: impact on genes regulating hepatic lipid metabolism. *Molecular Nutrition & Food Research* 54: S24–S30.

Ros, E. 2010. Health benefits of nut consumption. *Nutrients* 2: 652–682.

Sanchez, D., M. Quinones, B. Moulay, B. Muguera, M. Miguel, and A. Aleixandre. 2010. Changes in arterial blood pressure of a soluble cocoa fiber product in spontaneously hypertensive rats. *Journal of Agricultural and Food Chemistry* 58: 1493–1501.

Santos-Buelga, C., and A. Scalbert. 2000. Proanthocyanidins and tannin-like compounds- nature, occurrence, dietary intake and effects on nutrition and health. *Journal of the Science of Food and Agriculture* 80: 1094–1117.

Shen, C.-L., J.K. Yeh, J.J. Cao, and J.-S. Wang. 2009. Green tea and bone metabolism. *Nutrition Research* 29: 437–456.

Watanabe, M., and J. Ayugase. 2010. Effects of buckwheat sprouts on plasma and hepatic parameters in type 2 diabetic *db/db* mice. *Journal of Food Science* 75: H294–H299.

Yang, C.S., S. Sang, J.D. Lambert, and M.-J. Lee. 2008. Bioavailability issues in studying the health effects of plant polyphenolic compounds. *Molecular Nutrition & Food Research* 52: S139–S151.

Index

R.E. Aluko, *Functional Foods and Nutraceuticals*, Food Science Text Series,
DOI 10.1007/978-1-4614-3480-1, © Springer Science+Business Media, LLC 2012

Printed in the United States
By Bookmasters